T0339403

FUNDAMENTALS OF APPLIED RESERVOIR ENGINEERING

FUNDAMENTALS OF APPLIED RESERVOIR ENGINEERING

Appraisal, Economics, and Optimization

RICHARD WHEATON

Senior Lecturer at the University of Portsmouth, United Kingdom

Amsterdam • Boston • Heidelberg • London
New York • Oxford • Paris • San Diego
San Francisco • Singapore • Sydney • Tokyo
Gulf Proffessional Publishing is an imprint of Elsevier

G P
P

Gulf Professional Publishing is an imprint of Elsevier
50 Hampshire Street, 5th Floor, Cambridge, MA 02139, USA
The Boulevard, Langford Lane, Kidlington, Oxford, OX5 1GB, UK

Copyright © 2016 Elsevier Ltd. All rights reserved.

No part of this publication may be reproduced or transmitted in any form or by any means, electronic or mechanical, including photocopying, recording, or any information storage and retrieval system, without permission in writing from the publisher. Details on how to seek permission, further information about the Publisher's permissions policies and our arrangements with organizations such as the Copyright Clearance Center and the Copyright Licensing Agency, can be found at our website: www.elsevier.com/permissions.

This book and the individual contributions contained in it are protected under copyright by the Publisher (other than as may be noted herein).

Notices
Knowledge and best practice in this field are constantly changing. As new research and experience broaden our understanding, changes in research methods, professional practices, or medical treatment may become necessary.

Practitioners and researchers must always rely on their own experience and knowledge in evaluating and using any information, methods, compounds, or experiments described herein. In using such information or methods they should be mindful of their own safety and the safety of others, including parties for whom they have a professional responsibility.

To the fullest extent of the law, neither the Publisher nor the authors, contributors, or editors, assume any liability for any injury and/or damage to persons or property as a matter of products liability, negligence or otherwise, or from any use or operation of any methods, products, instructions, or ideas contained in the material herein.

British Library Cataloguing-in-Publication Data
A catalogue record for this book is available from the British Library

Library of Congress Cataloging-in-Publication Data
A catalog record for this book is available from the Library of Congress

ISBN: 978-0-08-101019-8

For information on all Gulf Professional Publishing visit our website at https://www.elsevier.com/

Publisher: Joe Hayton
Acquisition Editor: Katie Hammon
Editorial Project Manager: Kattie Washington
Production Project Manager: Sruthi Satheesh
Designer: Mark Roger

Typeset by TNQ Books and Journals

CONTENTS

Supplementary material of this book is available at
http://booksite.elsevier.com/9780081010198/

LIST OF FIGURES

FOREWORD

The aim of this textbook is to provide the fundamentals of reservoir engineering for BEng/BSc students in petroleum engineering and give an introduction to reservoir engineering for MSc students who are studying petroleum engineering for the first time. The book would also be useful to employees in other disciplines in the oil and gas industry who want to understand the basics of this important and central subject.

Modern reservoir engineering is very largely centered on numerical computer simulation, and a reservoir engineer in industry will spend much of her/his professional career building and running such simulators. High-powered computers now mean that geological interpretations consisting of many millions of grids cells can be used to build reservoir models, honoring the fundamental set of physical laws (conservation of mass, conservation of momentum, and thermodynamic laws) which will predict the movement of phases and hydrocarbon components and production through all stages of field life for any potential development scenario.

These are very powerful tools both in planning and optimizing developments and in monitoring field behavior once production commences. Because of this, reservoir engineering has moved ahead and is now a very different discipline from that of 30—40 years ago, when so much depended on analytical methods based on equations derived from the basic physical laws but needing numerous simplifying assumptions to be solvable.

Because of their power and ease of use, there are significant dangers in these numerical simulators and they are unfortunately often misused. There is a tendency within industry to construct very large simulation models, often with millions of grid cells, before first production (or even full appraisal). These are based on very limited data, and the results are almost meaningless. Modern simulators also come with very sophisticated "post-processor" software that provides very attractive and convincing production plots and three-dimensional representations of the reservoir. These have a strong influence on financial decisions at an early stage, and such decisions can often be difficult to reverse later.

The key for the practicing reservoir engineer is to be able to use models in an appropriate way, exercising good "engineering judgment," and to start the process for any field by using all available methods, including very simple numerical models, to begin to understand the basic "dynamics" of the reservoir—what are the major factors that will determine its performance? Large simulation models can come later, when we have a significant amount of historical production and other data. It is the aim of this textbook to encourage future reservoir engineers to use this approach.

A chapter specifically treating reservoir appraisal and development planning is included, as this will normally make up a large proportion of an engineer's activities. There is also a chapter on petroleum economics, since all decisions will ultimately depend on the economics and a reservoir engineer should understand the basics of this subject.

Unconventional resources (shale gas and oil, coal-seam gas, and heavy oil) are covered, as they will be a major part of the industry in future.

Excel software is provided, and many of the exercises depend on use of this. The idea is to provide students and other readers with a simple, easy-to-use tool for analysis of some basic field data. Exercises which in many books require long numerical calculations can now be carried out very effectively using such Excel spreadsheets.

There are appendices covering topics such as enhanced oil recovery, gas well testing, basic fluid thermodynamics, and mathematical operators, which are peripheral but should help in the understanding of the main topics.

The aim of this book is give a basis for an understanding of how hydrocarbon reservoirs work, and to start the process for a student developing "good reservoir engineering judgment."

CHAPTER 1

Introduction

The role of a reservoir engineer is a key and central one in petroleum engineering (Fig. 1.1). He/she pulls together all the available geological, petrophysical, laboratory, field, and well-test data to understand the physical potential of the reservoir. The engineer then covers the following aspects.

1. Reservoir evaluation.
2. Development planning and optimization.
3. Production forecasting.
4. Reserves estimation.
5. Building numerical reservoir models.
6. Well testing and analysis.
7. Field management.

To do this he/she also needs to understand the facilities and economic and commercial constraints, so as to provide and optimize a viable and economic development plan.

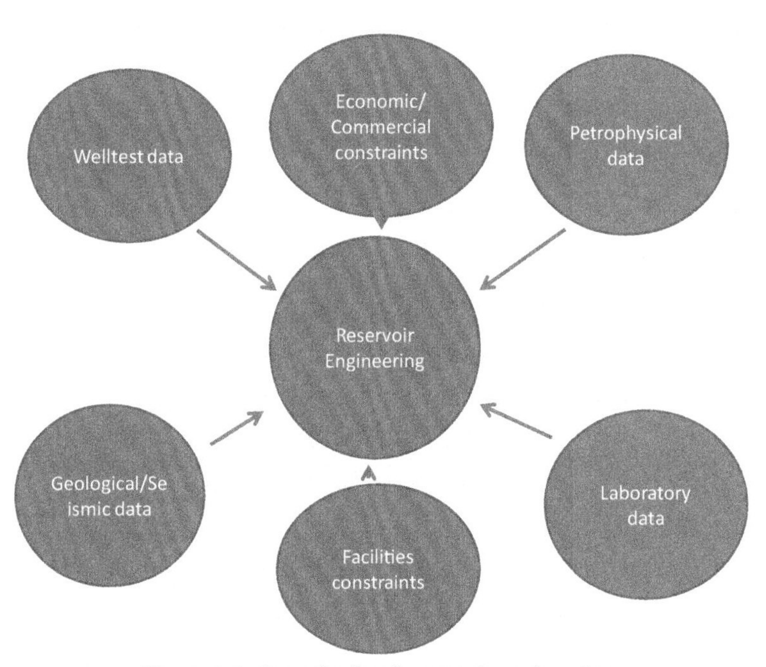

Figure 1.1 Central role of reservoir engineering.

Fundamentals of Applied Reservoir Engineering
ISBN 978-0-08-101019-8
http://dx.doi.org/10.1016/B978-0-08-101019-8.00001-6
© 2016 Elsevier Ltd.
All rights reserved.

To fulfill this role effectively, it is necessary for a reservoir engineer firstly to understand the basic physical properties relevant to reservoirs: the concepts of porosity, absolute permeability, wettability, capillary pressure, and relative permeability must be covered. Fluid properties then need to be understood: what hydrocarbon mixtures are typically found in fields, how these can split into oil and gas phases and how these phases behave with pressure and temperature. The reservoir engineer also needs to have a basic understanding of how all these properties are measured so that he/she can critically access the data he/she receives from the laboratory and the field. Chapter "Basic Rock and Fluid Properties" covers these fundamental issues.

Chapter "Well-Test Analysis" introduces well-test analysis, which, before production, provides our best insight into reservoir properties away from the very immediate vicinity of exploration and appraisal wells. The standard equations used are derived, and their interpretation explained. Software is provided to help in gaining experience on using these equations to interpret field-test data and also to answer exercises.

Analytical methods using simplified equations and models in the early evaluation of potential reservoir behavior are an important tool, covered in chapter "Analytical Methods for Prediction of Reservoir Performance". Material balance, mainly used for depletion-type developments, and Buckley-Leverett/Welge analysis for water-flood developments are discussed in some detail here. Again, software is available to aid in understanding these topics and in answering exercise questions.

Chapter "Numerical Simulation Methods for Predicting Reservoir Performance" gives an introduction to numerical simulation. The fundamental equations of mass balance, conservation of momentum (giving the Darcy equation), and thermodynamic relationships are combined to give the "diffusion equations," which are then solved across the grid cells of the simulator. The theory behind the use of finite difference methods is covered. The input data that will be required in any simulator are explained, and there is emphasis on the best use of numerical simulators. Use of production data as they become available in "history matching" to improve our model is discussed.

Once we understand the physical properties in the reservoir we need to consider the dynamics of the field when we drill wells and produce hydrocarbons. Pressure drops around the wells, and reservoir fluids move toward the well. Depending on the nature of the reservoir fluids, there will be some form of "drive mechanism" that maintains well production.

Production profiles and recovery of hydrocarbons will depend on the efficiency of this drive mechanism. In chapter "Estimation of Reserves and Drive Mechanisms" we consider the estimation of hydrocarbons in place and the drive mechanisms for all type of reservoir, and give the ranges of recovery factors that are typically achieved.

Chapter "Fundamentals of Petroleum Economics" gives the basics of petroleum economics. Decisions on developing a field will ultimately depend on economics. Reservoir engineers need to understand the economic indicators used to judge the value of a particular field development, how they are calculated and how they are used.

Spreadsheet software is provided with this publication that will input production profiles, expected gas or oil prices, discount rates, inflation rates, and taxation rates to give values for all the main economic indicators. It is intended that this software will be used in answering exercise questions.

Once fundamental reservoir properties, production drive mechanisms, and basic economic indicators are covered, we can look at reservoir appraisal and development planning – which is the topic of chapter "Field Appraisal and Development Planning". The appraisal and development planning stage is absolutely critical in obtaining value from an asset, and it is where reservoir engineers can have most influence on key decisions. Early decisions have the greatest financial impact on a project. This is known as "front-end loading." The appraisal and development planning process involves determining the critical sensitivities for a given reservoir (sensitivity analysis), what further data are needed to reduce uncertainty and risk (value of information analysis) and optimizing the development in terms of reservoir and necessary facilities. Software is provided to help in understanding this process and for use in exercise questions. Tools for use in early predictions, Analogue data and decline curve analysis are discussed in this chapter.

Unconventional resources are becoming increasingly important with the development of coal-bed methane and shale gas and oil. Initially exploited in the United States, they are now being developed worldwide. Chapter "Unconventional Resources" covers this topic, explaining the basic physics of these sources of hydrocarbons and the estimation of potential production profiles and reserves. In Chapter "Producing Field Management" we discuss the role of the reservoir engineer in production field management.

Companies, other than national oil companies, need to declare the reserves and resources that they hold so that investors can value them and their company shares. One responsibility of reservoir engineers is to provide reserves estimates for the fields they are working on. The final chapter (see chapter: Uncertainty and the Right to Claim Reserves) in this textbook examines the international rules of Society of Petroleum Engineers and Security and Exchange Commission (SPE and SEC) for levels of certainty on economically recoverable field reserves and resources. Probabilistic methods for estimating reserves are discussed.

Four supporting sections—fluid thermodynamics, gas well testing, enhanced oil recovery, and mathematical notes—are covered in appendices.

An understanding of the thermodynamics of multicomponent hydrocarbons, of why some mixtures split into gas and oil phases at certain pressure and temperature conditions, and how the volumes of these phases vary with pressure and temperature is useful in understanding the section on reservoir fluid behavior.

The use of mathematical calculus has been deliberately kept to a minimum in this text, and where it is necessary an attempt is made to explain the meaning of equations in the text. Such equations are not popular with some students, but they are the basis of the behavior of reservoirs and give a concise representation of the physical relationships involved. It is therefore worth to have a mathematical note (Appendix 2), where an attempt is made to clarify the significance of the various mathematical operators used in this and other reservoir engineering texts.

CHAPTER 2

Basic Rock and Fluid Properties

2.1 FUNDAMENTALS

There are four fundamental types of properties of a hydrocarbon reservoir that control its initial contents, behavior, production potential, and hence its reserves.

1. The rock properties of porosity, permeability, and compressibility, which are all dependent on solid grain/particle arrangements and packing.
2. The wettability properties, capillary pressure, phase saturation, and relative permeability, which are dependent on interfacial forces between the solid and the water and hydrocarbon phases.
3. The initial ingress of hydrocarbons into the reservoir trap and the thermodynamics of the resulting reservoir mixture composition.
4. Reservoir fluid properties, phase compositions, behavior of the phases with pressure, phase density, and viscosity.

In this chapter we look at the basics of each of these properties, and also consider how they are estimated.

2.2 POROSITY

2.2.1 Basics

Porous rock is the essential feature of hydrocarbon reservoirs. Oil or gas (or both) is generated from source layers, migrates upwards by displacing water and is trapped by overlying layers that will not allow hydrocarbons to move further upwards. Porous material in hydrocarbon reservoirs can be divided into clastics and carbonates. Clastics such as sandstone are composed of small grains normally deposited in riverbeds over long periods of time and covered and compressed over geological periods (Fig. 2.1). Carbonates (various calcium carbonate minerals) are typically generated by biological processes and again compressed by overlying material over long periods of time. Roughly 60% of conventional oil and gas resources occur in clastics and 40% in carbonates.

Fundamentals of Applied Reservoir Engineering
ISBN 978-0-08-101019-8
http://dx.doi.org/10.1016/B978-0-08-101019-8.00002-8
© 2016 Elsevier Ltd.
All rights reserved.

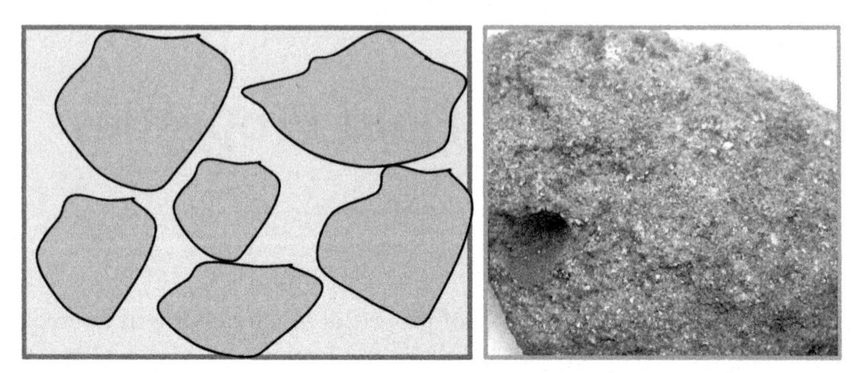

Figure 2.1 Solid grains making up porous rock (schematic image and sandstone photograph).

Porosity, given the symbol ϕ, is defined as the ratio of void volume to total rock volume:

ϕ = void volume/(grain + void volume)

$$\phi = \frac{V_p}{V_b} \quad [2.1]$$

where V_p = pore space volume and V_b = bulk volume (grain + void volume).

Porosity will depend on the average shape of the solid grains and the way they are packed together. This in turn will depend on the way the rock was formed from sedimentation over time—for example, solid grains of sand deposited gradually on riverbeds (clastics), or growth and decay of biological materials (carbonates). This initial distribution of solids is then often disturbed by subsequent events, which rearrange the solids' distribution, affecting the porosity (digenesis).

Typical porosity values are in the range 5–30%, with 15% as a very typical value.

In reservoir engineering we are normally only interested in interconnected porosity, which is the volume of connected pores to total bulk rock volume.

Hydrocarbon pore volume is the total reservoir volume that can be filled with hydrocarbons. It is given by the equation:

$$\text{HCPV} = V_b \cdot \phi \cdot (1 - S_{wc}) \quad [2.2]$$

where V = bulk rock reservoir volume and S_{wc} = connate (or irreducible) water saturation as a fraction of pore space.

As pressure decreases with hydrocarbon production, rock particles will tend to pack closer together so that porosity will decrease somewhat as a function of pressure. This is known as rock compressibility (c_r):

$$c_r = -\frac{1}{V_p}\frac{\partial V_p}{\partial P} \quad [2.3]$$

where $V_p = V_b \cdot \phi$ = pore volume.

Porosity of real rocks is often (in fact normally) very heterogeneous, depending on the lithology of the rock—which is typically variable even over quite short distances. Certainly, layer by layer and aerially within the reservoir porosity will vary significantly.

Also the geometry of the pore space is very variable, so that two samples of porous material, even if they have the same porosity, can have very different resistance to fluid flow. We discuss this in detail later.

2.2.2 Measurement of Porosity

Porosity is measured in two ways, from either wire line logs or laboratory measurement on core.

2.2.2.1 Wire Line Logs

Porosity can be estimated from interpretation of wire line logs, in particular acoustic, neutron, or gamma ray logs. Instruments are lowered down a well and measurements made and then interpreted to give reservoir porosity as a function of depth (Fig. 2.2).

2.2.2.2 Laboratory Measurement of Porosity

Porosity is calculated using the following equation:

$$\text{porosity} = \phi = \frac{V_p}{V_b} = \frac{V_b - V_m}{V_b} \quad [2.4]$$

where V_p = pore space volume, V_m = matrix (solid rock) volume, and V_b = bulk volume (= $V_p + V_m$).

We need two out of these three values to determine porosity.

Bulk volume (V_b) can be determined directly from core dimensions if we have a fluid-saturated regularly shaped core (normally cylindrical), or by fluid displacement methods—by weight where the density of the solid matrix and the displacing fluid is known, or directly by volume displacement.

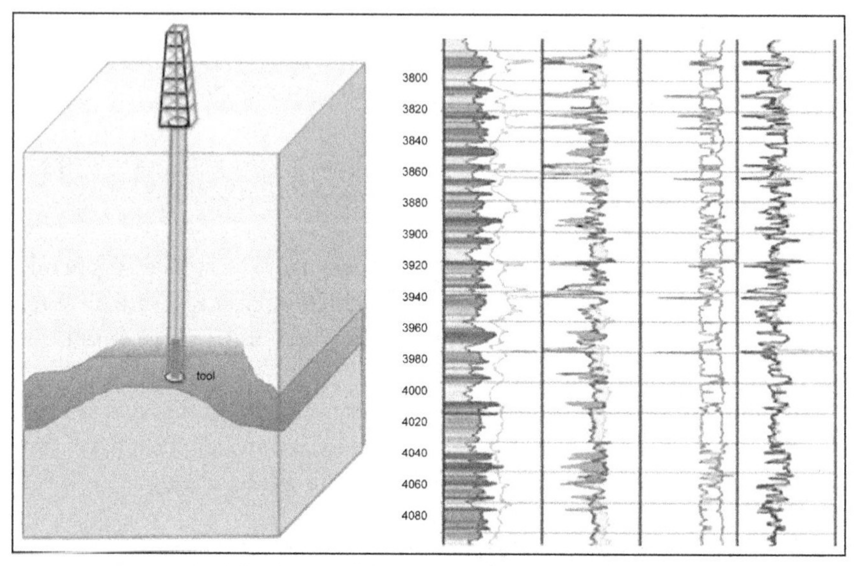

Figure 2.2 Wire line well logging—schematic and example log.

Matrix volume (V_m) can be calculated from the mass of a dry sample divided by the matrix density. It is also possible to crush the dry solid and measure its volume by displacement, but this will give total porosity rather than effective (interconnected) porosity. A gas expansion method can be used: gas in a cell at known pressure is allowed to expand into a second cell containing core where all gas present has been evacuated. The final (lower) pressure is then used to calculate the matrix volume present in the second cell using Boyle's law (Fig. 2.3). This method can be very accurate, especially for low-porosity rock.

Boyle's law: $P_1V_1 = P_2V_2$ (assuming gas deviation factor Z can be ignored at relatively low pressures) can now be used.

Pore space volume (V_p) can also be determined using gas expansion methods.

2.2.3 Variable Nature of Porosity

As discussed above, porosity is very variable in its nature, changing over quite small distances within a reservoir; and even if two samples have the same porosity, it does not mean that they will have the same absolute permeability or the same wettability characteristics, which in turn means that they can have very different capillary pressure and relative permeability

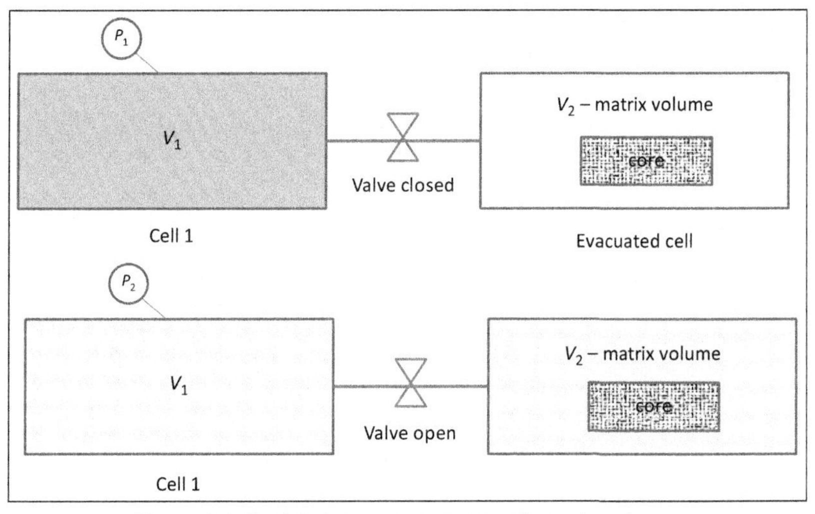

Figure 2.3 Boyle's law measurement of matrix volume.

properties. The key factors here are the average pore geometry and the polar/nonpolar nature of the rock itself.

The normal theoretical model for porous material is the "pore and throat" model, illustrated schematically in Fig. 2.4. In this model, the

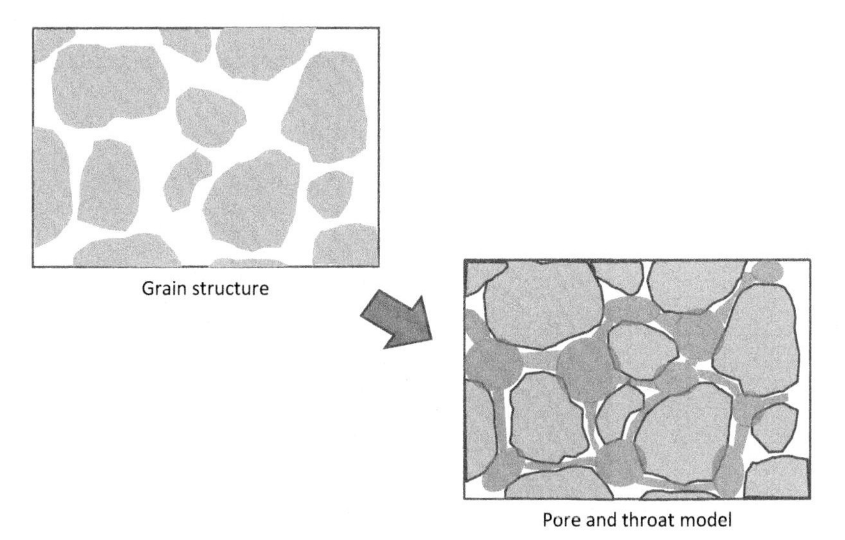

Figure 2.4 Pore and throat model.

volume is mainly in the "pores" and the flow characteristics will principally depend on the average geometry of the "throats."

Pore and throat geometry will depend on average grain size and shape at deposition (or on biological growth for carbonates) and on digenetic changes (ie, postdepositional rearrangements).

2.2.4 Net to Gross (NTG)

Some regions of a reservoir are often considered to have such low porosity and transport potential as to be effectively nonreservoir. They are therefore left out of estimated reservoir volume and considered as "dead" rock. So, for example, net thickness of a formation is = average gross thickness × NTG.

2.3 PERMEABILITY

2.3.1 Basics

Permeability is a key parameter in reservoir engineering. Darcy provided an empirical equation that related fluid flow through porous material to the pressure gradient and gravitation.

It is possible to derive the Darcy equation from first principles if various simplifying assumptions are made, and it is worthwhile understanding this.

From first principles: with *conservation of momentum* we can consider a volume element (V) of fluid moving through porous material. V will have a different position at time $t + \delta t$ compared with that at time t, and it may also have a different volume (Fig. 2.5).

The rate of change in momentum with time (from t to $t + \delta t$) equals body forces (gravity) + stress (frictional forces).

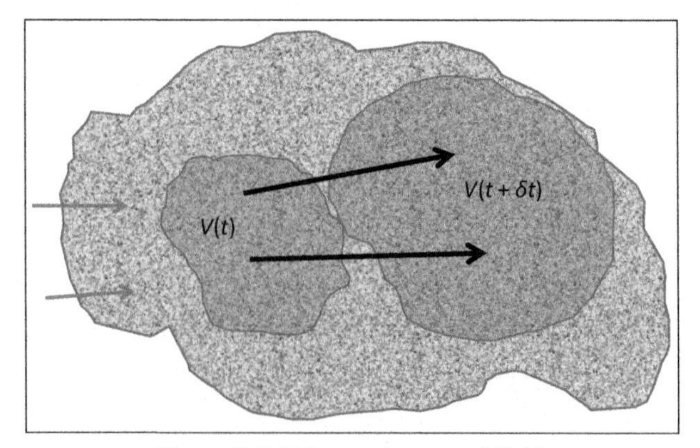

Figure 2.5 Volume element of fluid.

For a volume element, if we assume a steady state situation and ignore inertial effects (ie, flow rates are relatively small) we can derive the equation from conservation of momentum:

$$\nabla \bar{p}^V + \nabla \cdot \bar{\sigma}^V = \frac{1}{V_f} \int_{A_{fs}} \Psi s \cdot d\boldsymbol{A} + \int_V \rho \boldsymbol{F} dV \qquad [2.5]$$

See appendix "Mathematical Note" for an explanation of some the mathematical operators in this equation.

This equation simply represents a balance of average forces within a volume element, arising during the steady-state flow of a single-phase fluid through fractured material. The first term on the left-hand side represents the force due to any pressure gradient, while the second term represents frictional forces due to viscosity of the fluid. The first right-hand term describes frictional forces due to the solid rock matrix (ie, friction between moving fluid and the rock matrix), which will be much greater than those due to viscous effects within the fluid itself; we therefore neglect the second term on the left-hand side, while the second term on the right-hand side represents body forces (ie, gravity). We therefore obtain the following relationship:

$$\nabla \bar{p}^V = \frac{1}{V_f} \int_{A_{fs}} \Psi s \cdot d\boldsymbol{A} + \int_V \rho \boldsymbol{F} dV \qquad [2.6]$$

This can be simplified to

$$\nabla p = -\frac{\mu}{k} u + \rho g \nabla z \qquad [2.7]$$

where we assume that

$$k = K_g \overline{d^2} \phi \qquad [2.8]$$

where K_g is a geometric constant, $\bar{d}$ is an averaged "characteristic length" for the porous material, and ϕ = porosity. Rearranged, this gives the standard form of the **Darcy equation**:

$$\boldsymbol{u} = -\frac{k}{\mu} (\boldsymbol{\nabla} p - \rho g \boldsymbol{\nabla} z) \qquad [2.9]$$

2.3.2 Measurement of Permeability

2.3.2.1 Laboratory Determination of Permeability

Single-phase absolute permeability is measured on core in a steel cylinder where pressures P_1 and P_2 are measured for a given gas flow rate Q (Fig. 2.6).

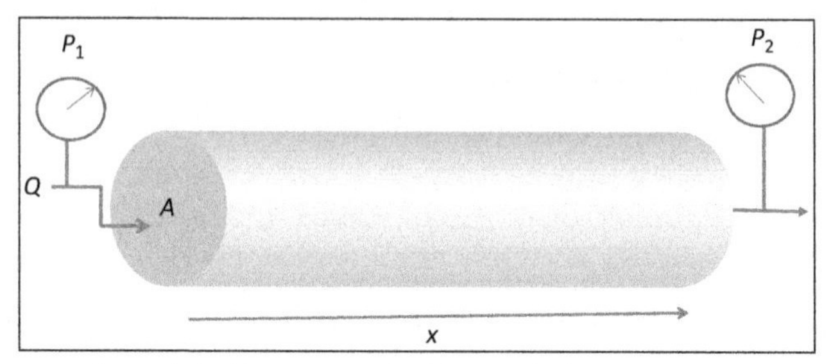

Figure 2.6 Measurement of permeability—schematic.

For a gas: from Darcy's law for horizontal flow,

$$Q = \frac{kA\left(P_1^2 - P_2^2\right)}{2\mu x} \quad [2.10]$$

For an incompressible liquid: for horizontal flow,

$$Q = \frac{kA(P_1 - P_2)}{\mu x} \quad [2.11]$$

where Q = volumetric flow rate (cm^3/s); A = area (cm^2); μ = viscosity of the gas or liquid; P = pressure (atmospheres); x = length of core (cm). This gives the value for permeability k in Darcy's equation.

2.3.2.2 Permeability From Well-Test Analysis

For a constant production flow rate Q, permeability can be estimated from average formation thickness h, fluid viscosity μ, bottom hole pressure P_w, initial reservoir pressure P_e at an assumed undisturbed (still at initial conditions) distance r_e from the well and wellbore radius r_w using the above equations. This is discussed further in chapter "Field Appraisal and Development Planning."

$$Q = \frac{2\Pi kh(P_e - P_w)}{\mu \ln\left(\frac{r_e}{r_w}\right)} \quad [2.12]$$

Units are as above; see Fig. 2.7(a).

2.3.2.3 Darcy's Law in Field Units

In field units the Darcy equation will be

$$u = -1.127 \times 10^{-3} \frac{k}{\mu}\left[\frac{dp}{dx} + 0.4335\Upsilon \sin \alpha\right] \quad [2.13]$$

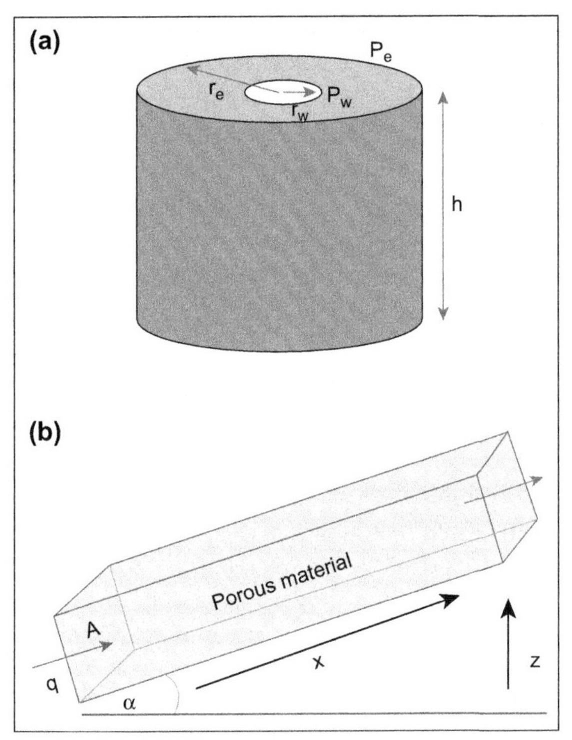

Figure 2.7 Permeability plots. (a) Radial coordinates. (b) Field unit dimensions.

where k is in milliDarcies (mD); u is in RB/day/ft^2; $\frac{dp}{dx}$ is in psi/ft; μ is in centipoise (cP); and Υ is specific gravity (dimensionless); see Fig. 2.7(b).

2.3.3 Permeability Variation in a Reservoir

Permeability measured from core is obviously very local depending on the nature of the porous rock, but as discussed above this changes continually across a reservoir depending on depositional and subsequent rearrangement effects. A reservoir can be divided into what are called "flow units"—regions which have common permeability, porosity, and wettability (and hence flow) characteristics. We saw above that

$$k = K_G \overline{d^2} \phi \qquad [2.14]$$

The permeability of each flow unit will depend on the average characteristic length (d), geometric constant (K_G), and porosity (ϕ) in that unit.

The characteristic length d arises from the shear stress relationship at solid surfaces due to the velocity gradient. It thus represents an averaged distance between the channel-centered velocity point and the stationary rock surfaces—ie, the averaged flow channel radius for pores and throats.

The geometric constant K_G has two components, both related to the "averaged" geometry of the channels through which the fluid flows. The first arises from the nature of the averaged shear stress in the fluid, which is strongly dependent on average throat diameter: the smaller the average throat diameter the larger the shear stress and the larger is this component. The second comes from the averaged relationship between volume and surface area, essentially the ratio of pore (spherical) volume to channel (cylindrical-type) volume.

The simplifying assumption made here is that for single-phase flow, "channel" geometry is essentially independent of porosity for different rock types.

If we assume that we have a set of hydraulic unit-type systems with common geometric features, and that $K_G \cdot d^2$ is approximately a constant for each with respect to porosity, we will have a range of permeability–porosity relationships like those shown in Fig. 2.8.

This is generally in line with most experimental porosity permeability data.

In reservoir modeling, assumptions will have to be made on the distribution of flow units across a reservoir. Analysis of well-test data can give valuable information on permeability distribution across significant regions of a reservoir away from the wells.

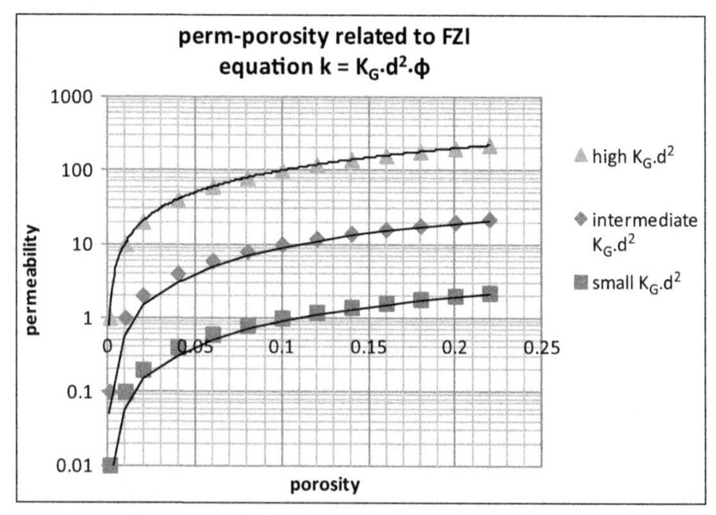

Figure 2.8 Porosity–permeability relationships.

2.3.4 Vertical and Horizontal Permeability

It is normally (but not always) assumed that horizontal permeability is the same in each direction; but vertical permeability is often, and particularly in clastics, significantly smaller than horizontal permeability when sediments are frequently poorly sorted, angular, and irregular. Vertical/horizontal (k_v/k_h) values are typically in the range 0.01–0.1.

2.4 WETTABILITY

2.4.1 Basics

When two immiscible fluids are in contact with a solid surface, one will tend to spread over or adhere to the solid more than the other. This is the result of a balance of intermolecular forces and surface energies between fluids and the solid. This is shown in Fig. 2.9, where vector

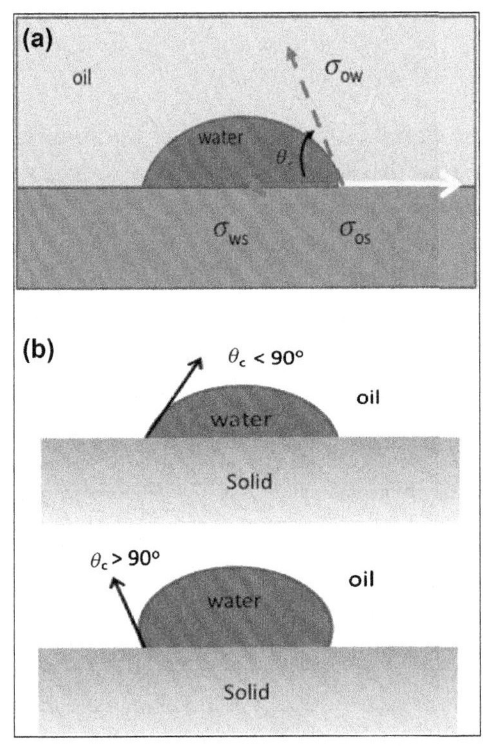

Figure 2.9 (a) Water–oil–solid interfacial interactions. (b) Contact angles. Where $\theta_c < 90°$, the system is known as "water wet" and water will tend to spread on the solid surface; and where $\theta_c > 90°$, the system is known as "oil wet" and oil will spread on the solid surface.

forces are balanced at the oil–water–solid contact point, giving the relationship

$$\sigma_{os} - \sigma_{ws} = \sigma_{ow} \cos \theta_c \quad [2.15]$$

where σ_{os} = the interfacial tension between oil and solid; σ_{ws} = the interfacial tension between water and solid; σ_{ow} = the interfacial tension between oil and water; and θ_c = the contact angle between water and oil at the contact point measured through the water.

Wettability will control the distribution of oil and water in the pore space. In water wet systems oil will tend to be found in the centers of pores, while in oil wet systems oil will be retained around the solid grains (see Fig. 2.10). This will of course have a fundamental effect on oil recovery in water flooding. Many examples of porous material have intermediate wettability where the contact angle is close to 90°. We can also have short-range variable or "mixed" wettability.

Gas will normally be thc nonwetting phase with respect to both water and oil.

2.4.1.1 Hysteresis

The history of the porous rock (in terms of the history of the phases—water–oil or gas—that have occupied the pore space) will have a strong effect on its wettability; this is known as "hysteresis."

Wettability is fundamental in determining capillary pressure and relative permeability (discussed later).

2.4.1.2 Imbibition and Drainage

Imbibition is the phenomenon of increasing wetting-phase occupation of pore space, while drainage is a decrease in the wetting phase present.

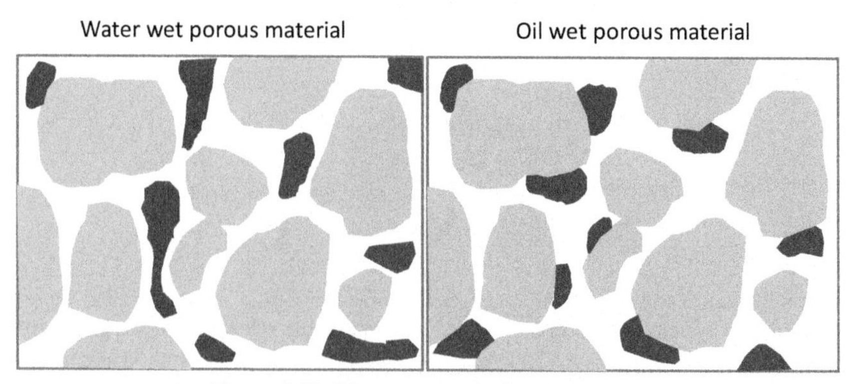

Figure 2.10 Water wet and oil wet systems.

2.4.2 Measuring Wettability

Several methods are available to measure a reservoir's wetting preference. Core measurements include imbibition and centrifuge capillary pressure measurements (discussed below). An **Amott imbibition test** compares the spontaneous imbibition of oil and water to the total saturation change obtained by flooding. We will also see later that capillary pressure and relative permeability measurements give an idea of rock wettability.

2.5 SATURATION AND CAPILLARY PRESSURE

2.5.1 Saturation

Saturation is the proportion of interconnected pore space occupied by a given phase. For a gas—oil—water system,

$$S_w + S_o + S_g = 1 \quad [2.16]$$

where S_w = water saturation, S_o = oil saturation, and S_g = gas saturation.

2.5.2 Capillary Pressure

Capillary pressure is the average pressure difference existing across the interfaces between two immiscible fluids, so for an oil—water system,

$$P_{cow} = p_o - p_w \quad [2.17]$$

It will depend on the average water/oil/rock contact angle (θ) and the average pore space radius (r).

Therefore capillary pressure is a function of both average wettability and average pore **size** (see Fig. 2.11).

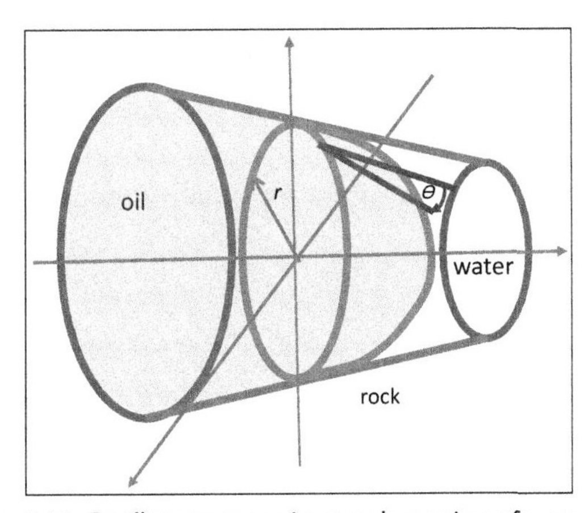

Figure 2.11 Capillary pressure in sample section of pore space.

If we take an example section of pore space, it can be shown that the capillary pressure between the wetting and nonwetting phases is given by

$$P_{\text{cnw}} = 2 \cdot K_{\text{G}} \sigma_{\text{nw}} \cos(\theta)/r_{\text{w}} \quad [2.18]$$

where K_{G} is a geometric constant dependent on the average geometry of the pore space, σ_{nw} is the oil/water interfacial tension, and r_{w} is the average radius of wetting-phase-occupied pores.

This wetting-phase radius r_{w} is a function of the saturation of the wetting phase. The smaller the wetting-phase saturation the more it will be concentrated in smaller pores, so that r_{w} will be small and hence the capillary pressure will be larger. The smaller the interfacial tension between the two phases the smaller will be the capillary pressure.

A drainage (decreasing wetting phase) capillary pressure curve is shown below (see Fig. 2.12(b)). Pb is the threshold pressure—the minimum pressure required to initiate drainage displacement. Capillary pressure increases as wetting phase saturation decreases, since

$$P_{\text{cap}} = 2K_{\text{G}}\sigma \cos(\theta)/r_{\text{w}} \quad [2.19]$$

and it can be shown that

$$k \propto \bar{r}^2 \varphi = K_{\text{G}}^* \bar{r}^2 \phi$$

where K_{G}^* is another geological constant. Thus,

$$\bar{r} \propto \sqrt{k/\phi} \quad [2.20]$$

and

$$\bar{r}_{\text{w}} \propto \bar{r} \left(S_{\text{w}}^{\text{N}}\right)^{\text{an}} \quad [2.21]$$

so that,

$$P_{\text{cap}} = \frac{K_{\text{G}}^* \sigma \cos\theta}{\sqrt{\frac{k}{\phi}} \left(S_{\text{w}}^{\text{N}}\right)^{\text{aw}}} \quad [2.22]$$

Leverett's J function (a dimensionless capillary pressure) is defined as

$$J(S_{\text{w}}) = \frac{P_{\text{cap}}}{\sigma \cos\theta} \sqrt{\frac{k}{\phi}} = \frac{K_{\text{G}}^*}{\left(S_{\text{w}}^{\text{N}}\right)^{\text{aw}}} \quad [2.23]$$

where $S_{\text{w}}^{\text{N}} = \frac{S_{\text{w}} - S_{\text{wc}}}{1 - S_{\text{wc}} - S_{\text{or}}}$

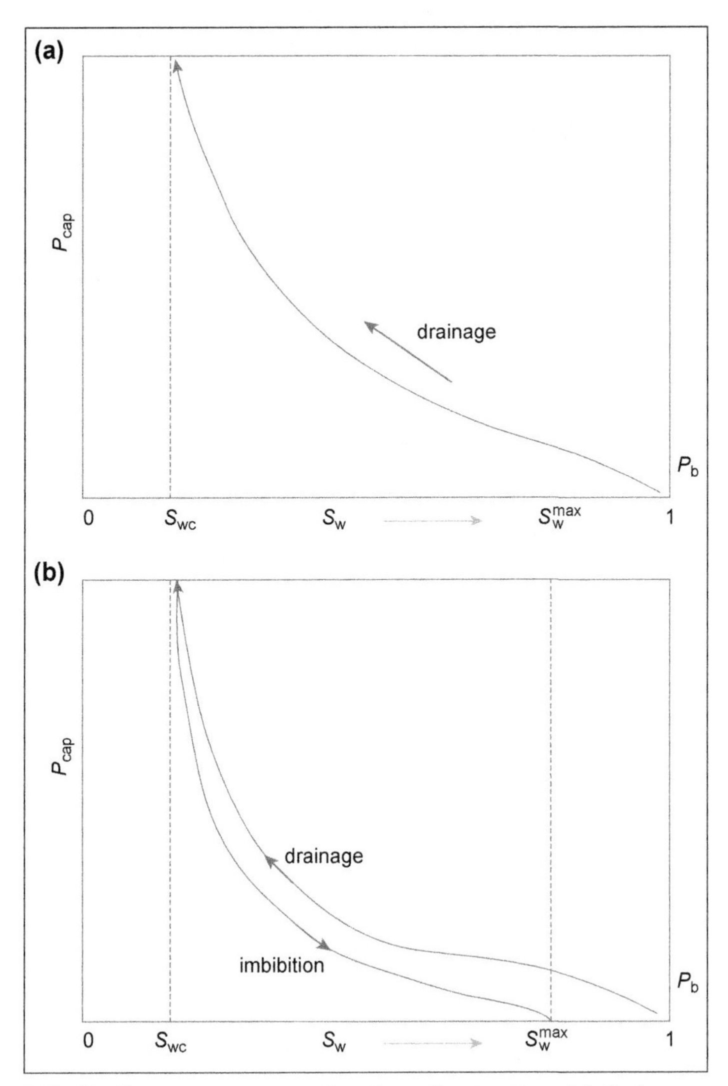

Figure 2.12 Capillary pressure as a function of saturation. (a) Drainage capillary pressure curve. (b) Drainage and imbibition capillary pressure curves.

This is a laboratory-measured relationship for a drainage experiment. Leveretts's *J* function has been used to characterize rock types, but it is difficult to use and unreliable.

Fig. 2.12 shows a comparison of drainage with an imbibition-type experiment.

Oil/water capillary pressure is of more significance than oil/gas capillary pressure, which is normally very small and can often be neglected.

2.5.3 Reservoir Saturation With Depth

The major importance of capillary pressure is its effect on the distribution of phases in the reservoir with depth.

For each phase k, reservoir pressure increases with depth (z) depending on phase density:

$$\frac{dP_k}{dz} = \rho_k g \qquad [2.24]$$

and since,

$$P_o - P_w = P_{cow}$$

$$\frac{dP_{cow}}{dz} = -(\rho_o - \rho_w) \cdot g \qquad [2.25]$$

so

$$\frac{\Delta P_{cow}}{\Delta z} = -(\rho_o - \rho_w) \cdot g \qquad [2.26]$$

if $\Delta z =$ width of transition zone, where

$$\Delta P_{cow} = P_{cow}(S_o = 1 - S_{wc}) - P_{cow}(S_o = 0)$$

but

$$P_{cow}(S_o = 0) = 0$$

so that

$$\Delta z = -\frac{P_{cow}(S_o = 1 - S_{wc})}{(\rho_o - \rho_w) \cdot g} \qquad [2.27]$$

High permeability or contact angle (close to 90°) results in small capillary pressures, thus in this case we have a smaller transition zone. Low permeability or small contact angle systems (with large capillary pressures) will have wide transition zones.

Fig. 2.13 shows a schematic for an oil—water reservoir of oil and water pressures as a function of depth (left-hand plot) and of oil/water capillary pressure as a function of water saturation (right-hand plot).

Geologically, the reservoir will have initially been filled with water ($S_w = 100\%$). Oil migrating up from below (oil having a lower density than water) will have gradually displaced water—*a drainage process* (see Fig. 2.14).

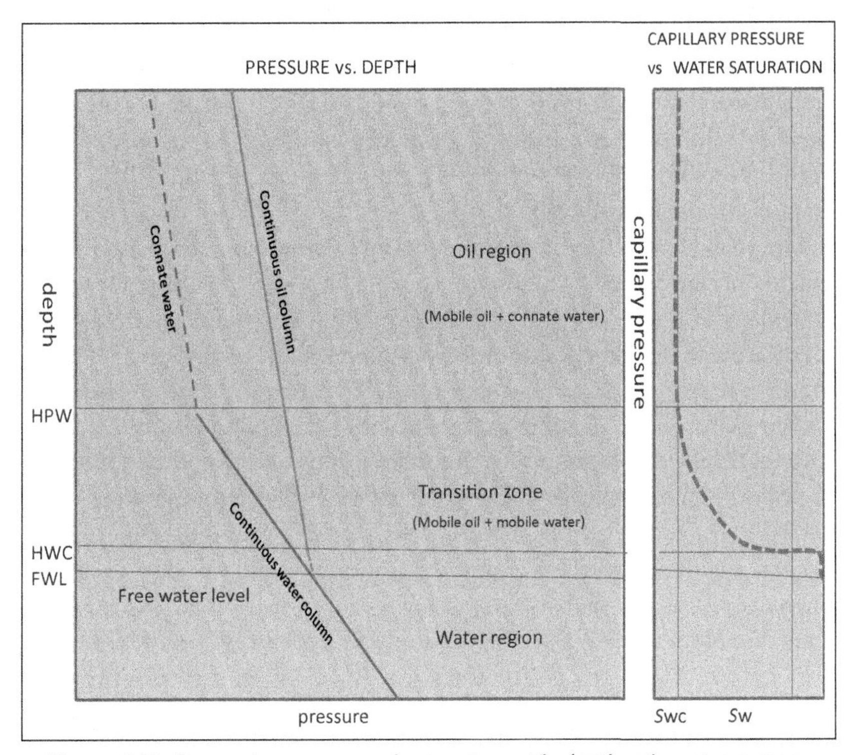

Figure 2.13 Reservoir pressure and saturation with depth: oil–water system.

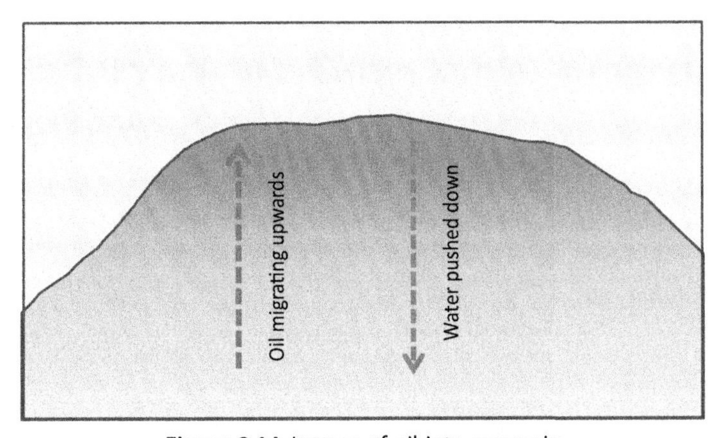

Figure 2.14 Ingress of oil into reservoir.

In Fig. 2.13 we have a free water level (FWL), defined as the depth at which the aquifer water pressure gradient and the hydrocarbon pressure gradient meet; the hydrocarbon water contact (HWC), defined as the depth below which no hydrocarbon is produced but in which discontinuous hydrocarbon may exist; and a highest produced water depth (HPW), above which no water is produced.

The thickness of the transition zone will depend on the nature of the capillary pressure curve, as shown above. This in turn depends on the distribution of pore space sizes in the rock. High-permeability rocks with a preponderance of larger well-connected pores will normally have a shallow capillary pressure curve and a correspondingly narrow transition zone, like that shown in Fig. 2.15, in comparison with the above plot.

In real systems the situation is almost always more complicated than this. We will have a number of "rock types" or rock units (see discussion under wettability), which each have their own capillary pressure characteristics. Also we can have perched aquifers in layered reservoirs due to the fill history.

As discussed above, capillary pressure will depend on permeability and contact angle (Fig. 2.16), so the thickness of the oil—water transition zone will depend on permeability and the oil—water contact angle.

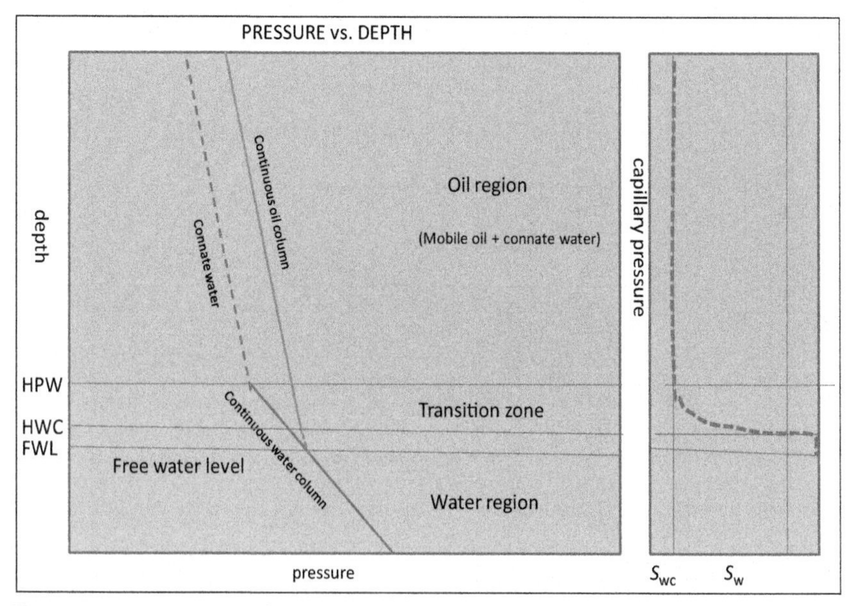

Figure 2.15 Pressure and saturation with depth—shallow capillary pressure curve (small transition zone).

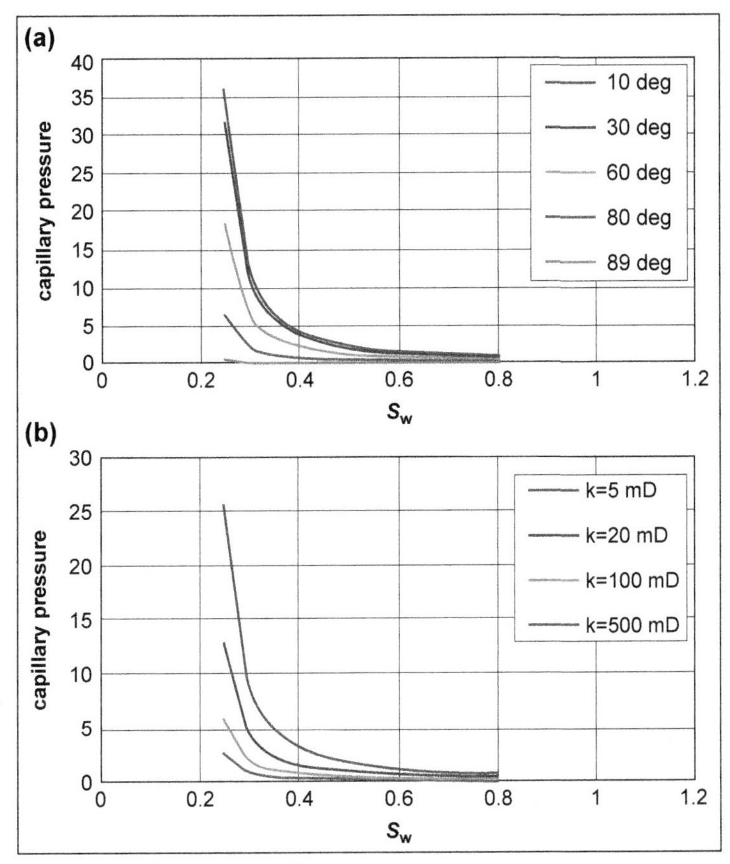

Figure 2.16 Dependence of capillary pressure curves on (a) contact angle and (b) permeability.

2.5.3.1 Oil–Water Reservoirs With a Gas Cap

These reservoirs have a situation like that shown in Fig. 2.17, with a gas–oil contact depth. The gas–oil transition zone is normally so narrow that it can be neglected.

2.6 RELATIVE PERMEABILITY

2.6.1 Basics

If we have more than one fluid phase flowing simultaneously through a porous medium, each has its own effective permeability that will depend on the saturation of each fluid:

$$k_e = k \cdot k_r \qquad [2.28]$$

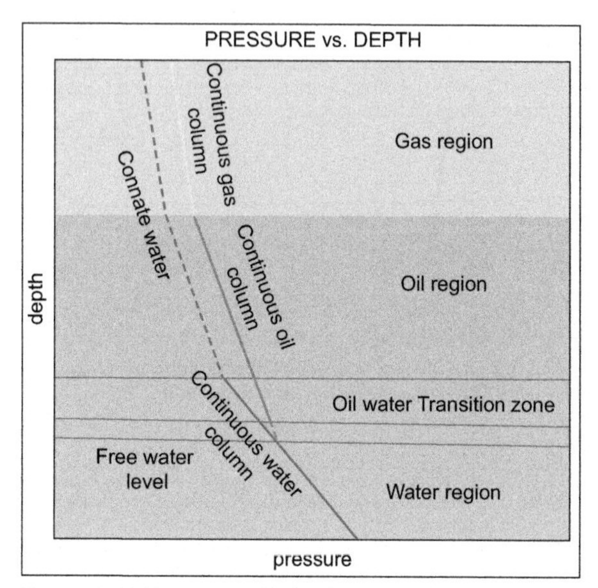

Figure 2.17 Pressure with depth: gas–oil–water system.

where k_e = effective permeability, k = absolute permeability, and k_r = relative permeability.

Darcy's law then becomes

$$u_\alpha = -\frac{kk_{r\alpha}}{\mu_\alpha}\left(\frac{dp}{dx} + \rho_\alpha g\frac{dz}{dx}\right) \qquad [2.29]$$

for phase α, where $k_{r\alpha} = f(S_\alpha)$.

Relative permeability is a very important parameter in reservoir engineering, but it is unfortunately a very difficult function to measure in the laboratory and then to relate to what is likely to happen in the reservoir.

2.6.2 Oil–Water Systems

We first look at a two-phase oil–water system, a typical plot for which is shown in Fig. 2.18(a).

At connate (or irreducible) water saturation (S_{wc}), oil relative permeability is at its maximum (k_{rom}). As water saturation increases (imbibition), oil relative permeability decreases and water relative permeability increases until no more oil can be displaced by water, at which point oil saturation = S_{or} (irreducible oil saturation) and water saturation $S_w = 1 - S_{or}$. At this point water relative permeability is at a maximum (k_{rwm}). The broken lines outside this region from $S_w = S_{wc}$ to $S_w = 1 - S_{or}$ are not

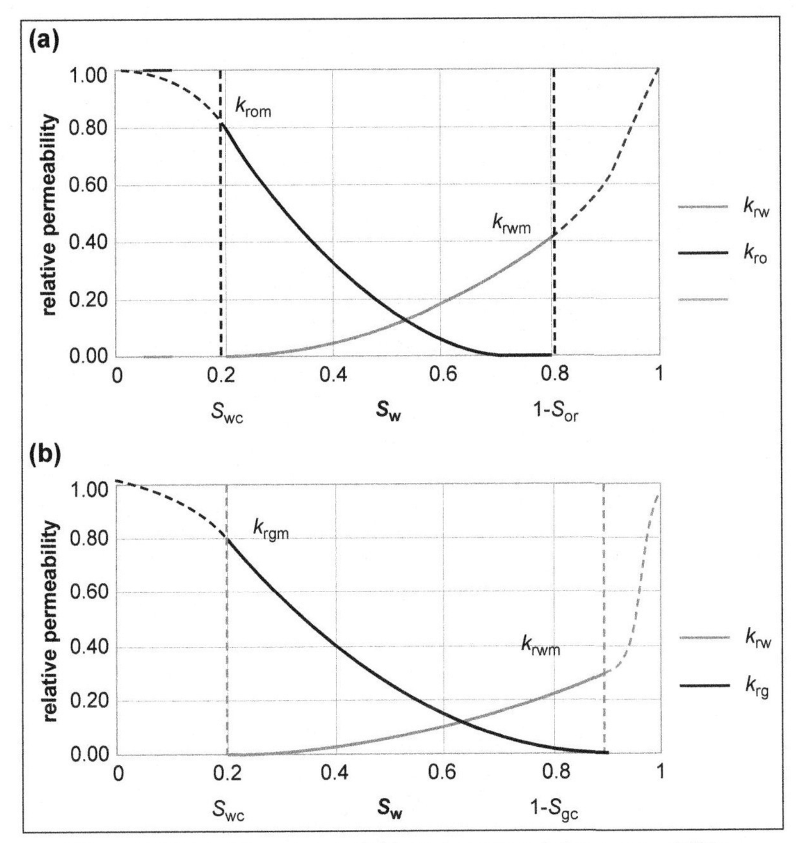

Figure 2.18 (a) Oil—water and (b) gas—water relative permeabilities.

reached in the reservoir, but can correspond to some laboratory experiments.

The situation in the reservoir during production will normally be imbibition, as water from either an aquifer or water injection displaces oil, driving it toward the producing wells. Because of the heterogeneous nature of reservoirs, however, there will in some areas be a limited drainage situation. There is a hysteresis effect, but this is relatively small for straight oil—water systems.

The crossover point for the oil and water relative permeability curves is often indicative of the water wet or oil wet nature of the porous material. Where the crossover occurs with $S_w < 0.5$ we will assume an oil wet system. When crossover occurs with $S_w > 0.5$ this would normally be considered a water wet system (see Fig. 2.19).

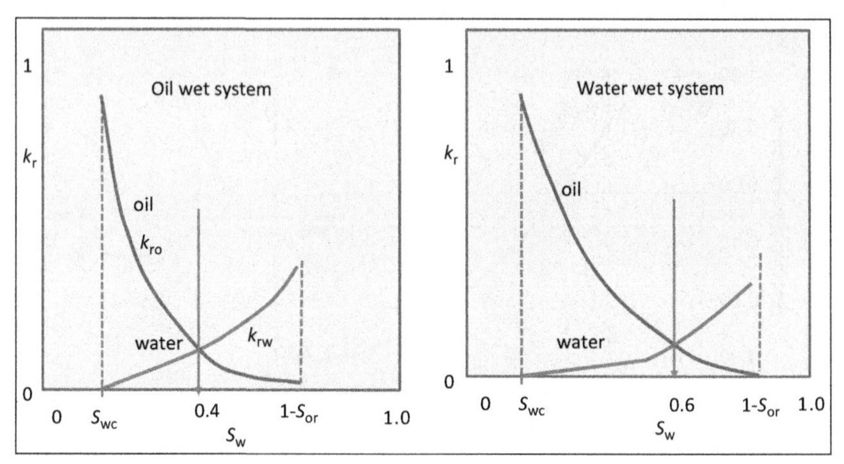

Figure 2.19 Oil wet and water wet systems.

2.6.3 Gas—Water Systems

In these systems gas will be the nonwetting phase, so plots will be comparable to the oil—water plots (Fig. 2.18(b)), with gas replacing oil.

At S_{wc} we have a maximum gas relative permeability of k_{rgm}, and at minimum or residual gas saturation S_{gc}, we have a maximum water relative permeability of k_{rwm}.

2.6.4 Gas—Oil Relative Permeability

In this case oil is the wetting phase and gas the nonwetting phase (see Fig. 2.20(a)).

There are two experimental situations: gas—oil measurements with connate water present and oil—water measurements with residual gas present (see Fig. 2.20(b)).

At $S_o = S_{oc}$ (irreducible oil saturation), maximum gas relative permeability is k_{rgm}, while at irreducible gas saturation S_{gc} maximum oil relative permeability is k_{rom}.

With oil—gas relative permeability, the molar compositions of the oil and gas will affect the interfacial tension between the two phases, which changes S_{oc} and S_{gc} and also the shape of the curves. As gas—oil interfacial tension decreases, both residual oil and gas saturations will decrease to zero in the limit interfacial tension going to zero (since the two phases are now indistinguishable). Gas injection into oil will result in changes in the composition of both oil and gas, which will reduce S_{oc} and S_{gc}.

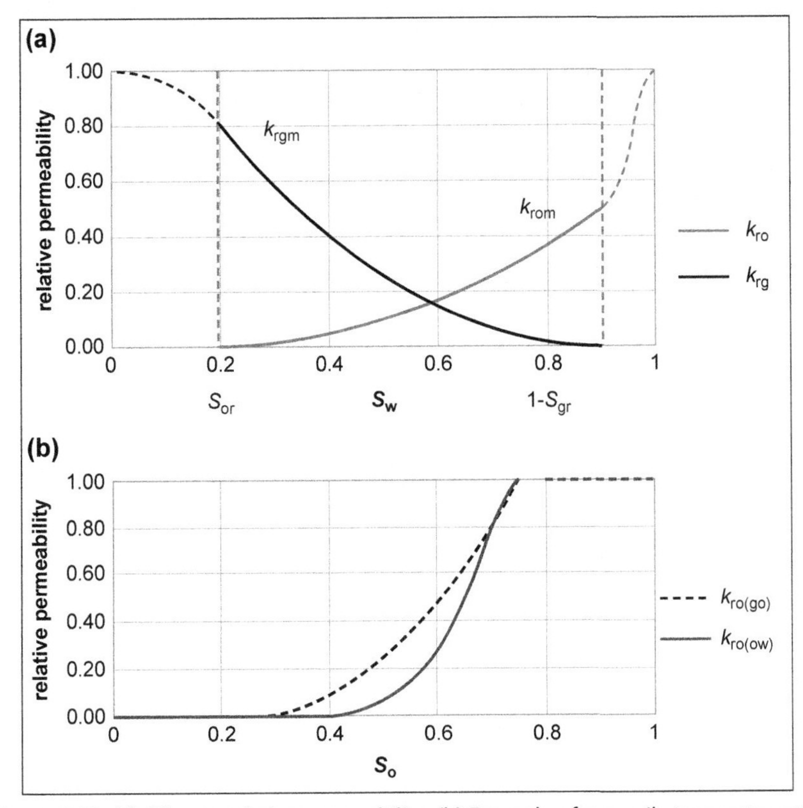

Figure 2.20 (a) Oil–gas relative permeability. (b) Example of gas–oil at connate water and oil–water at residual gas.

There are often significant hysteresis effects depending on the saturation direction—either drainage or imbibition in the wetting phase. These are not covered in this textbook, and the reader is referred to the "Further Reading" list at the end of this chapter.

2.6.5 Semi-Empirical Equations for Two-Phase Relative Permeabilities

Relative permeabilities are laboratory-measured parameters (Table 2.1), but where these measurements are not yet available, for studies of hypothetical reservoirs and as a tool for increased understanding of the effects of variations in relative permeability on reservoir reserves and production, semi-empirical equations such as those below are valuable. These are used in available Excel software.

Table 2.1 Range of relative permeability parameter values

Range	k_{ro}^{max}	k_{rw}^{max}	k_{rg}^{max}	S_{wc}	S_{or}	S_{gc}	n_o	n_w	n_g
Maximum	0.90	0.60	0.2	0.3	0.4	0.10	3.0	5.0	1.5
Minimum	0.80	0.30	0.6	0.1	0.2	0.20	2.0	2.0	2.5
Average	0.85	0.45	0.4	0.2	0.3	0.15	2.5	3.0	2.0

Oil relative permeability (oil—water systems):

$$k_{ro} = k_{ro}^{max}\left(\frac{1 - S_w - S_{oc}}{1 - S_{wc} - S_{oc}}\right)^{n_o}$$

Water relative permeability (oil—water systems):

$$k_{rw} = k_{rw}^{max}\left(\frac{S_w - S_{wc}}{1 - S_{wc} - S_{oc}}\right)^{n_w}$$

Gas relative permeability (gas—water systems):

$$k_{rg} = k_{rg}^{max}\left(\frac{S_g - S_{gc}}{1 - S_{wc} - S_{gc}}\right)^{n_g}$$

2.6.6 Three-Phase Relative Permeabilities

This is a more complex situation and occurs in oil fields where oil goes below its bubble-point pressure, so that gas is released, or where there is gas ingress from a gas cap, or where there is gas injection in an oil—water system. It also occurs in gas condensate systems where we go below the dew point.

The system is more difficult and complex not only because we now have three phases—gas, oil, and water present in a volume of porous material—but because the history of that volume of rock (in terms of which phases have previously been in the pore space) is critical in determining the relative permeabilities of each phase.

A number of theoretical models have been proposed relating three-phase relative permeabilities to two-phase plus residual third-phase relative permeabilities.

The relationship between three-phase and two-phase relative permeabilities is illustrated in a ternary saturation schematic (Fig. 2.21(a)). The two-phase measurements discussed above (oil—water with no gas present

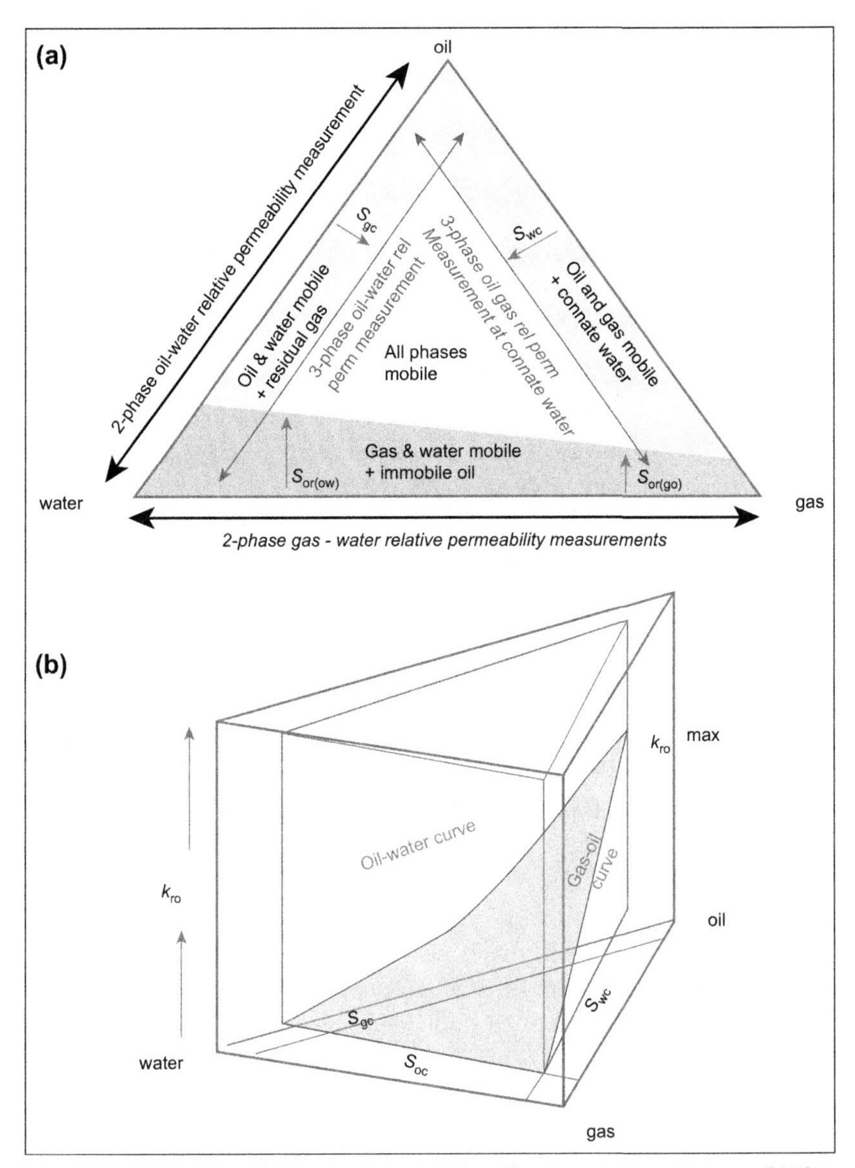

Figure 2.21 (a) Ternary oil–gas–water saturation diagram at given pressure. (b) Three phase oil relative permeability.

and gas–water with no oil present) are shown by the black arrows in the diagram. The three-phase relative permeabilities commonly measured (oil–water with residual gas and oil–gas with connate water) are shown by red arrows.

So, for example, oil-phase relative permeability in a three-phase system can be represented as shown in Fig. 2.21(b).

There are a number of relationships that interpolate the relative permeability of oil into a three-mobile-phase region from two-phase oil—water + residual gas and gas—oil + connate water systems (red arrows—gray in print versions in Fig. 2.21). An example of this is

$$k_{ro} = \frac{S_o^N k_{ro}(S_o, S_{wc}) \cdot k_{ro}(S_{gc}, S_o)}{k_{ro}^{max} \cdot \left(1 - S_g^N\right)\left(1 - S_w^N\right)} \quad [2.30]$$

where S_o^N, S_g^N, and S_w^N are normalized oil, gas, and water saturations.

Empirical equations for the two oil relative permeability relationships (oil—gas with connate water and oil—water with residual gas) are given below. An example plot is shown in Fig. 2.22.

$$k_{ro}(S_o, S_{wc}) = k_{ro}^{max}\left(\frac{1 - S_g - S_{wc} - S_{oc}}{1 - S_{wc} - S_{oc} - S_{gc}}\right)^{n_{o(go)}} \quad [2.31]$$

$$k_{ro}(S_o, S_{gc}) = k_{ro}^{max}\left(\frac{1 - S_w - S_{gc} - S_{oc}}{1 - S_{wc} - S_{oc} - S_{gc}}\right)^{n_{o(ow)}} \quad [2.32]$$

2.6.7 Measurement of Relative Permeability

There are two ways of measuring relative permeabilities in the laboratory.

1. Steady-state methods.
2. Unsteady-state methods.

Steady-state methods involve the simultaneous injection of two or more phases into a core of porous material. The flow ratio is fixed, and the test proceeds until an equilibrium is reached such that the pressure drop across the core has stabilized. The data obtained are used with Darcy's law to calculate the relative permeabilities of each phase. The flow ratio is changed to give relative permeabilities over the full range of saturations.

The advantage of steady-state methods is that it is simple to interpret resulting data. It is, however, time-consuming since a steady state can take many hours to achieve.

Unsteady-state methods are an indirect technique in which the relative permeabilities are determined from the results of a simple displacement test. Flow-rate data for each phase are obtained from the point at which the injected phase breaks through and we have two-phase flow. Unsteady-state

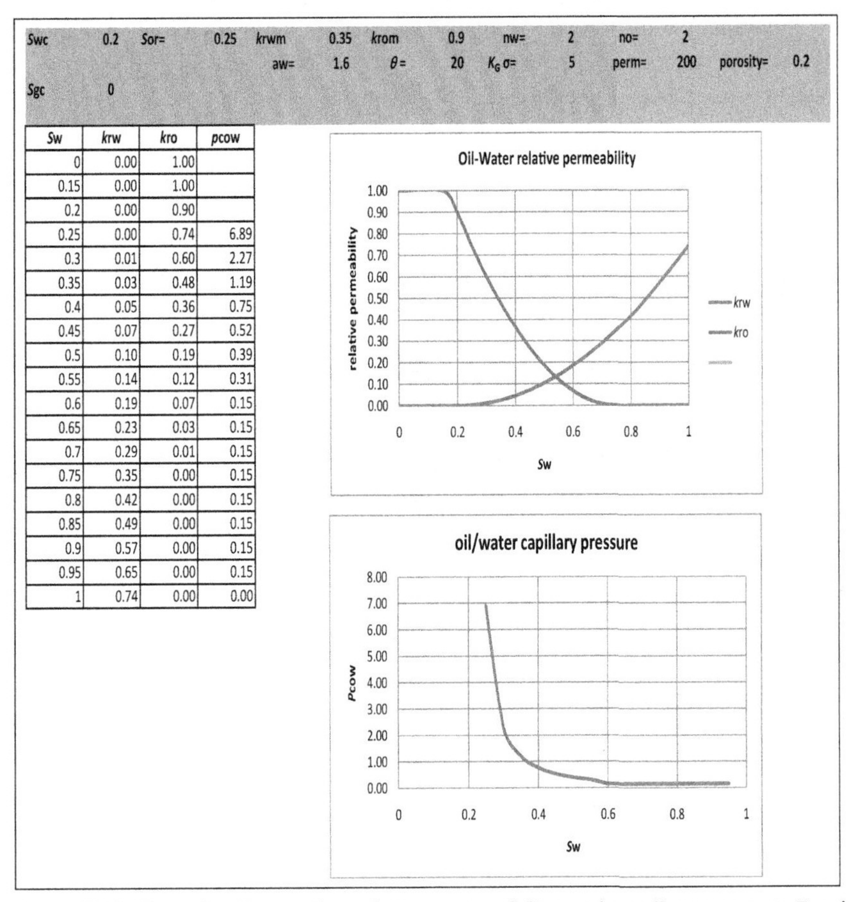

Sw	krw	kro	pcow
0	0.00	1.00	
0.15	0.00	1.00	
0.2	0.00	0.90	
0.25	0.00	0.74	6.89
0.3	0.01	0.60	2.27
0.35	0.03	0.48	1.19
0.4	0.05	0.36	0.75
0.45	0.07	0.27	0.52
0.5	0.10	0.19	0.39
0.55	0.14	0.12	0.31
0.6	0.19	0.07	0.15
0.65	0.23	0.03	0.15
0.7	0.29	0.01	0.15
0.75	0.35	0.00	0.15
0.8	0.42	0.00	0.15
0.85	0.49	0.00	0.15
0.9	0.57	0.00	0.15
0.95	0.65	0.00	0.15
1	0.74	0.00	0.00

Figure 2.22 Example of use of a relative permeability and capillary pressure Excel spreadsheet.

methods have the advantage of being quick to carry out, but data are much more difficult to interpret.

2.6.8 Excel Software for Producing Empirical Relative Permeability and Capillary Pressure Curves

A spreadsheet on "relative permeability and capillary pressures" is available where the semi-empirical equations discussed above are used to generate relative permeability and capillary pressure curves. An example of its use is shown in Fig. 2.22 (input parameters in green cells).

The effect of the various input parameters on two-phase oil–water and gas–water curves and on oil–water with residual gas and oil–gas with connate water can be examined.

2.7 RESERVOIR FLUIDS

2.7.1 Basics

Reservoir fluids are a complex mixture of many hundreds of hydrocarbon components plus a number of nonhydrocarbons (referred to as inerts).

We will be considering

- phase behavior of hydrocarbon mixtures;
- dynamics of reservoir behavior and production methods as a function of fluid type—volumetrics; and
- laboratory investigation of reservoir fluids.

Reservoirs contain a mixture of hydrocarbons and inerts.

Hydrocarbons will be C_1 to C_n where $n > 200$.

The main inerts are carbon dioxide (CO_2), nitrogen (N_2), and hydrogen sulphide (H_2S).

Hydrocarbons are generated in "source rock" by the breakdown of organic material at high temperature and pressure, then migrate upwards into "traps" where permeable rock above displaces the water originally present (see Fig. 2.23).

The fluid properties of any particular mixture will depend on reservoir temperature and pressure.

The nature of the hydrocarbon mixture generated will depend on the original biological material present, the temperature of the source rock and the pressure, temperature, and time taken.

A number of phases of migration can occur, with different inputs mixing in the reservoir trap. In the reservoir we can eventually have single-phase (unsaturated) or two-phase (saturated) systems.

2.7.1.1 Hydrocarbons

A few examples of the hydrocarbons commonly occurring are shown in Fig. 2.24. Methane, ethane, and propane are always present in varying amounts (dominating in gases); normal and isobutane and pentane are also normally present. C6+ (up to C200 or more) will dominate in oils.

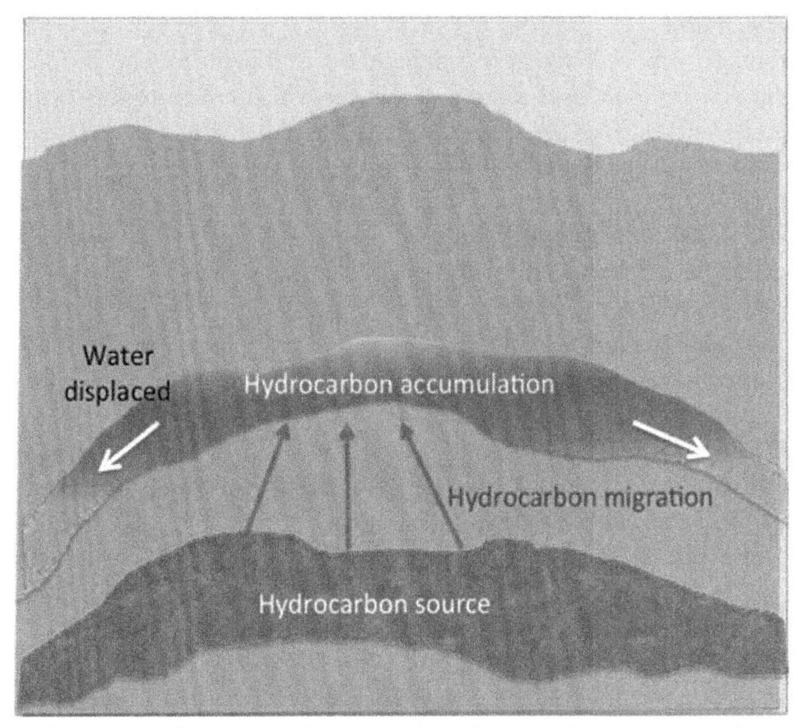

Figure 2.23 Migration and accumulation of hydrocarbons in a reservoir.

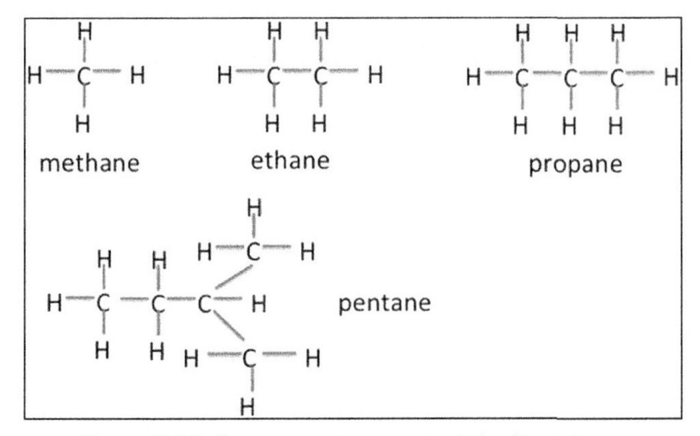

Figure 2.24 Some common reservoir hydrocarbons.

2.7.1.2 Inerts

Carbon dioxide and hydrogen sulphide are a problem for the petroleum engineer—they give acid solutions in water which are corrosive to metal pipelines and wellbore pipes. We also have the cost of removal, and in some H_2S cases even the disposal of unwanted sulfur is a problem.

2.7.1.3 Types of Reservoir Fluid

There are five types of reservoir fluid.

- Dry gas reservoirs.
- Wet gas reservoirs.
- Gas condensate.
- Volatile oil.
- Heavy oil.

Which fluid is present will depend on the total hydrocarbon mixture composition and its pressure and temperature.

Some typical properties of these reservoir fluid types are shown in Table 2.2. The methane (C1) molar fraction will typically be above 90% for a dry gas and below 60% for heavy (black) oil. The C5+ content will be negligible in a dry gas and more than 30% in heavy oil. API (American Petroleum Institute) is a measure of density ($^{\circ}API = 141.5/Sg_{60} - 131.5$ relates this to specific gravity relative to water at 60°F). Gas–oil ratio (GOR) is the gas content at 1 atm pressure and 60°F.

The range of temperatures and pressures to be considered by the reservoir engineer needs to cover those found in reservoirs through to atmospheric conditions with all possible temperature and pressures in between which may be encountered in the wellbore, surface pipelines, and separators (see Fig. 2.25).

Table 2.2 Range of reservoir fluid properties

	Dry gas	Wet gas	Gas condensate	Volatile oil	Black oil
C1	>0.9	0.75–0.90	0.70–0.75	0.60–0.65	<0.60
C2–4	0.09	0.10	0.15	0.19	0.11
C5+	–	–	0.1	0.15–0.20	>0.30
API	–	<50	50–75	40–50	<40
GOR (scf/stb)	–	>10,000	8000–10,000	3000–6000	<3000

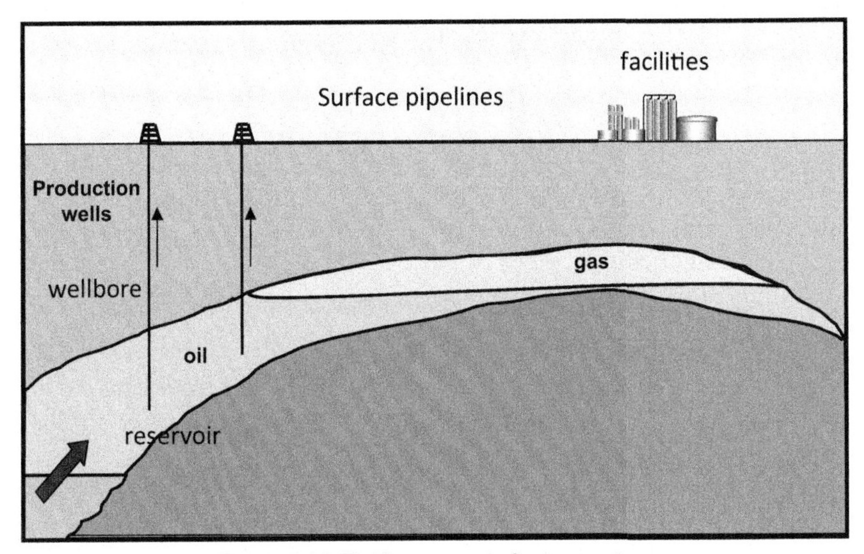

Figure 2.25 Fluid property reference points.

Reservoir temperature will depend on depth and the regional or local geothermal gradient. Reservoirs are found at depths between 1500 and 13,000 ft and a typical value of the geothermal gradient is 0.016°F/ft, so, for example, a reservoir at 5000 ft may have a temperature of 80°F and values between 50°F and 120°F are common.

We would typically expect to have a hydrostatic pressure gradient of ~0.433 psi/ft, which would correspond to reservoir pressures between 600 and 6000 psi. However, the hydrostatic gradient can be significantly more than this, and reservoir pressures in excess of 7000 psi are common.

Typical total molar content of the various reservoir fluid types are shown in Fig. 2.26.

There are two factors that determine the behavior of a reservoir containing any of these types of fluid as pressure and temperature change.

1. Fractional split into gas and oil phases, and composition of these phases.
2. Volume dependence on pressure and temperature of the two phases.

The first of these depends on thermodynamics—what is the most favorable state that minimizes free energy? The second depends on intermolecular forces. A detailed study of these factors is given in appendix "Basic Fluid Thermodynamics," but here we cover the resulting fluid behavior.

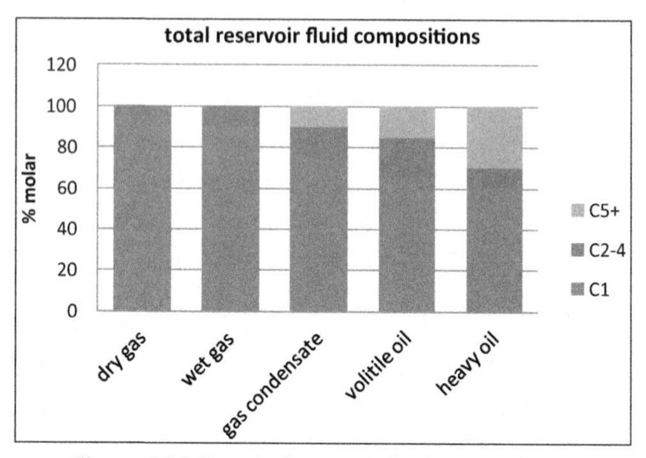

Figure 2.26 Range of reservoir fluid compositions.

2.7.2 Relationship Between Gas and Oil Phases—Single-Component Systems

In a single-component system, the component exists as a single phase at any temperature and pressure, as shown in Fig. 2.27. Below the critical point (at critical pressure and temperature) there is a pressure-temperature combination at which we move directly from liquid to gas or gas to liquid. An example is water, where at a pressure of 15 psi and a temperature of 100°C water boils, and as we input heat energy, liquid water is converted to steam.

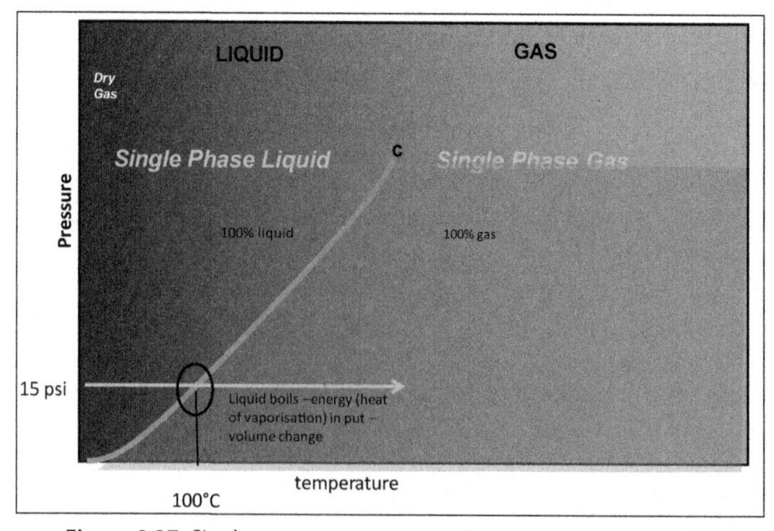

Figure 2.27 Single-component pressure/temperature relationship.

The temperature will remain constant at 100°C until all the water is in the gaseous phase.

2.7.3 Phase Equilibria in Multicomponent Systems

In multicomponent systems such as those that occur in hydrocarbon reservoirs, a similar phase diagram will look like that shown in Fig. 2.28. Rather

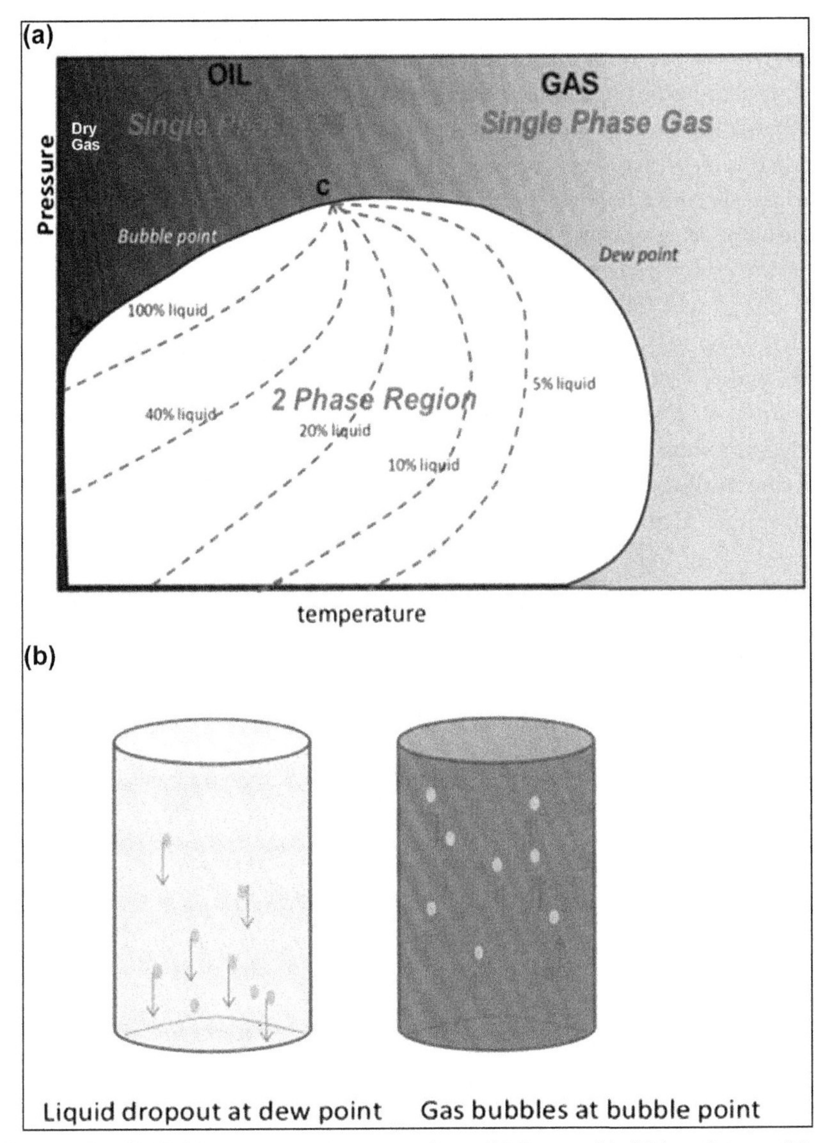

Figure 2.28 (a) Multicomponent phase envelope. (b) Dew and bubble point condition.

than a line division between gas and liquid, we now have a two-phase region where various proportions of the gas and oil phases coexist.

This represents the *pressure/temperature (PT) two-phase envelope* for a particular hydrocarbon mixture. At the temperatures and pressures in the green areas we have single-phase systems with this mixture. To the right of the critical point (C) we have a gas, so that if we drop the pressure until we cross the dew-point line a liquid drops out; and to the left an oil, where dropping the pressure to bubble-point pressure bubbles of gas appear. However, in the white area this mixture cannot exist as a single phase and will spontaneously split into a two-phase gas and oil system (see appendix: Mathematical Note). The percentage of liquid is shown by the broken lines inside this two-phase region. We can see that if temperature is kept constant and the pressure is reduced, the percentage of liquid increases before eventually decreasing again.

If we consider the reservoir fluid types discussed above, the shapes of the two phases will follow the pattern shown in Fig. 2.29.

It is important to understand the behavior of each of these reservoir types with PT changes both in the reservoir and between the reservoir and the surface. This is shown in Fig. 2.30, where the blue line follows decreasing pressure (with production) with constant temperature (normally the case within the reservoir itself), and the decrease in both pressure and temperature as we move to surface conditions.

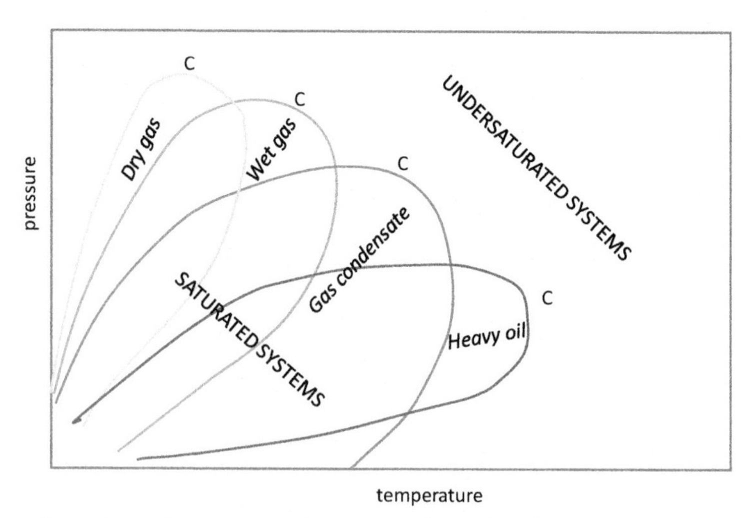

Figure 2.29 Comparison of phase envelopes for different fluid types.

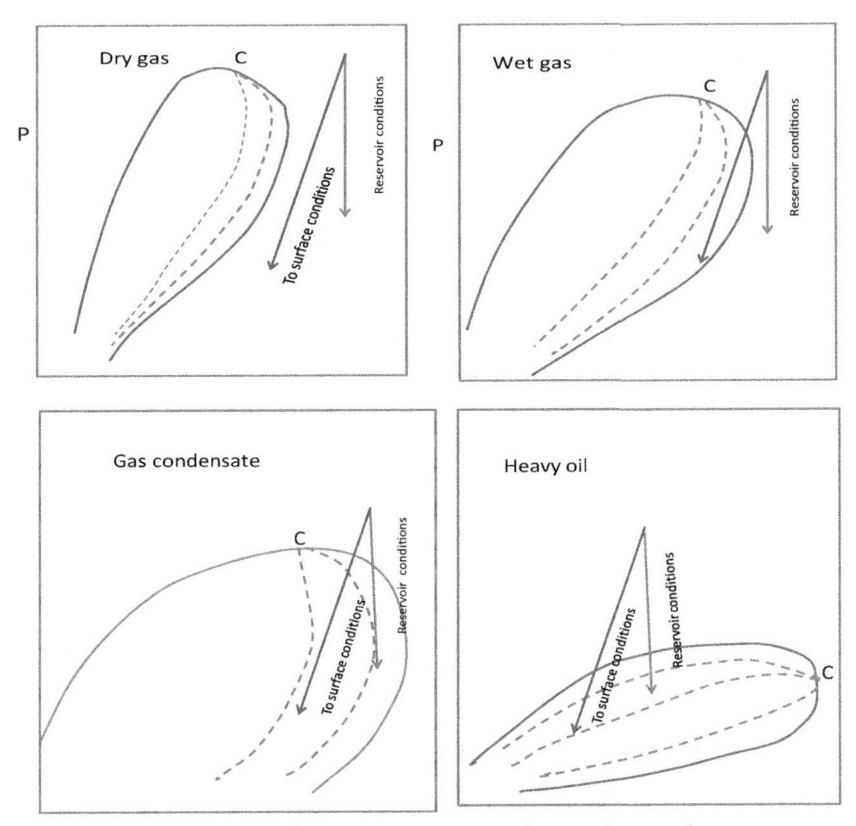

Figure 2.30 PT changes in reservoir and reservoir to surface.

For a dry gas we remain in the single-gas-phase region both in the reservoir and in the wellbore and at the surface. With a wet gas system we stay in the gas phase in the reservoir (so there is no liquid dropout), but the decrease in temperature between reservoir and surface means that we get liquid dropout potential in the wellbore, the pipelines, or the surface separator facilities.

In gas-condensate reservoirs we start above the dew point, and with decreasing pressure, liquid drops out within the reservoir, as well as liquid coexisting with gas in the wellbore, in the surface pipelines and being separated in the surface facilities.

For oil reservoirs when we go below the bubble point, both in the reservoir and in the wellbore, gas is evolved.

2.7.3.1 A Different Representation—Two-Pseudocomponent Pressure Composition Plots

The above PT plots are standard representation for pressure/volume/temperature (PVT) properties, but are often misleading—particularly for the reservoir. In this case temperature is normally more or less constant anyway. It must be remembered that the above PT plots are for a *fixed hydrocarbon mixture* composition.

When a reservoir is produced and goes below the dew point or bubble point, gas and oil move and at different rates—so, for example, gas is produced and the reservoir mixture changes, at least around the well. Although a two-pseudocomponent pressure composition representation is a gross simplification of the real system, it is much better for some uses. Let us assume, therefore, that we can represent the true reservoir mixture with just two pseudocomponents: a C1−C4 component and a C5+ component. A pressure composition plot will then have the general form shown in Fig. 2.31.

The vertical axis represents pressure, while the horizontal axis gives the fractions of C1−4 and C5+. On the far left we have C1−4 alone (where we have a gas phase only), and on the far right-hand side C5+ only (liquid

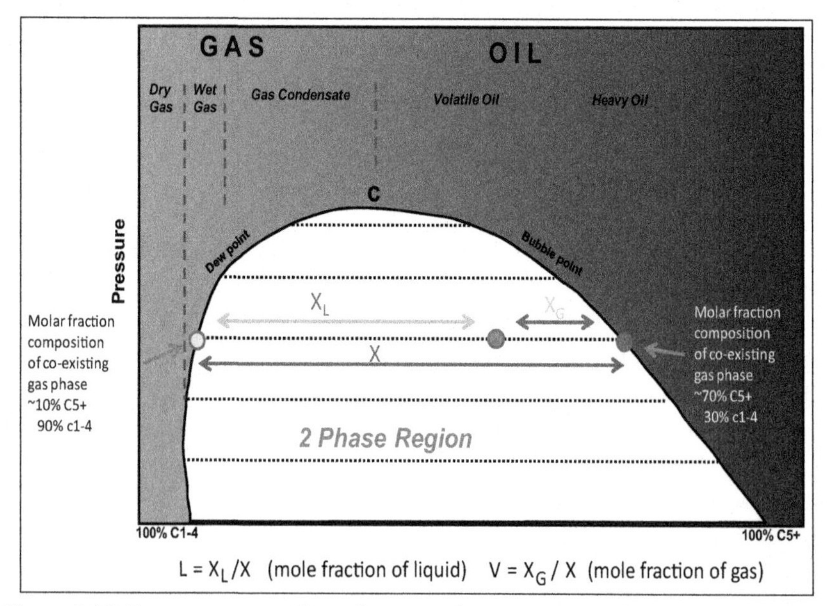

Figure 2.31 Pressure-composition phase envelope—two-pseudocomponent system.

at all pressures). The areas in green represent a single-phase region—gas to the left of the critical point and liquid to the right. In the white region compositions cannot exist as single phases at these pressures, but split into a gas and a liquid whose compositions correspond to the points at the ends of the tie lines (*dotted lines* in the figure).

In this example, at the pressure shown the gas will have a mole fraction of ~90% C1−4 and 10% C5+. The coexisting liquid phase has 70% C5+ and 30% C1−4. The number of moles of each phase is given by the fractions $L = X_L/X$ and $V = X_V/X$. The green two-phase region is split into regions where decreasing pressure gives dry gas, wet gas, gas condensate, volatile and heavy oil.

Another useful representation is the use of ternary diagrams, like that shown in Fig. 2.32.

This uses three pseudocomponents (C1, C2−4, and C5+) rather than two, so gives a rather better representation of the real system. The split of a reservoir mixture in the two-phase region between liquid and gas phases is shown in Fig. 2.32. We then need to consider how the two-phase envelope changes with pressure. We will not cover this rather complex subject any further here.

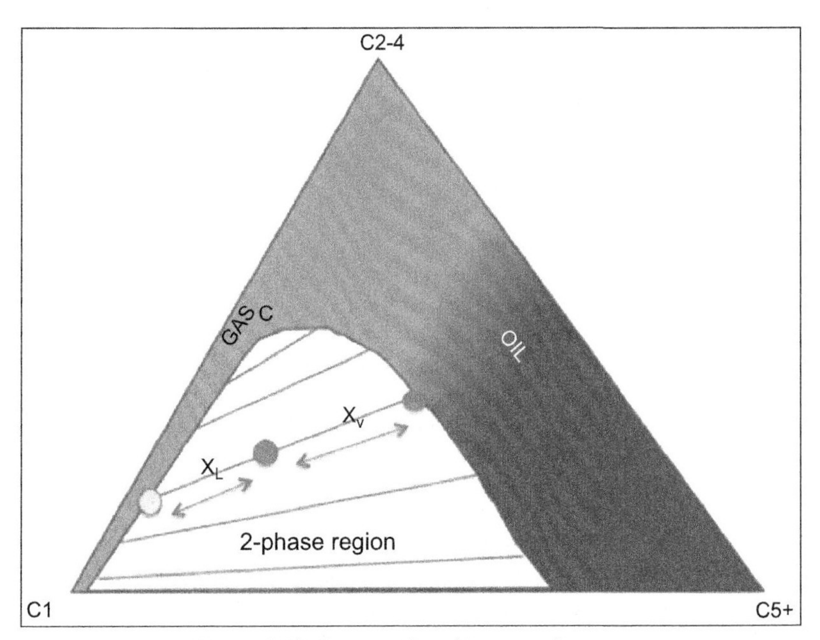

Figure 2.32 Compositional ternary diagram.

2.7.4 Volume Changes With Pressure and Temperature (PVT Relationships)

For single-component systems, the relationship of volume to pressure and temperature is shown in Fig. 2.33.

If we look at the pressure–volume curve at a temperature **above the critical temperature** (T_c) (curve NP), increasing pressure gives a single-phase system at all pressures.

At a temperature below the critical temperature (curve ABDE in the figure), as we increase pressure between points E and D we are in single phase; at D liquification starts, and volume decreases rapidly at this pressure until we have 100% liquid at B. Further increase in pressure gives only a small decrease in volume. The critical temperature and pressure are at point C. Above the critical temperature no pressure increase will give a liquid. Close to this point gas and liquid are very similar.

For a multicomponent system, the position is similar, except that as we go from liquid to gas expansion is accompanied by a decrease in pressure (see Fig. 2.34).

2.7.5 Obtaining Representative Reservoir Fluid Samples

Reservoir fluids are sampled in two ways to obtain representative samples of the original reservoir fluid.

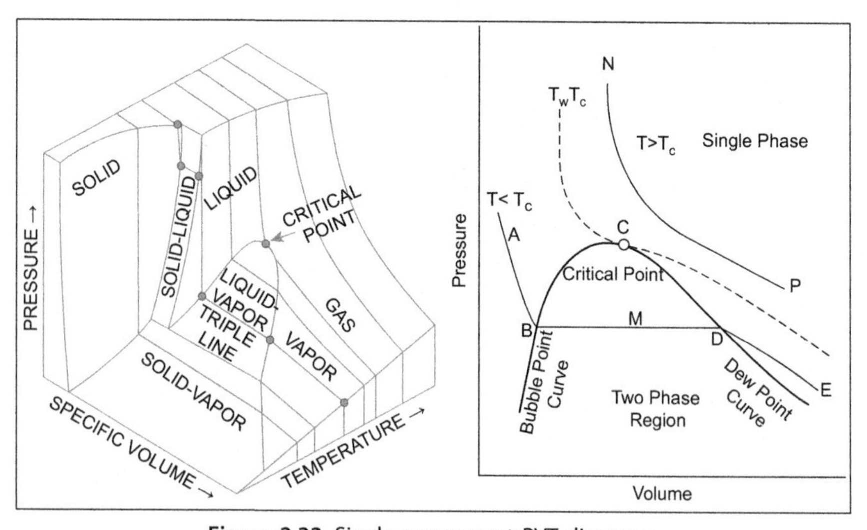

Figure 2.33 Single-component PVT diagrams.

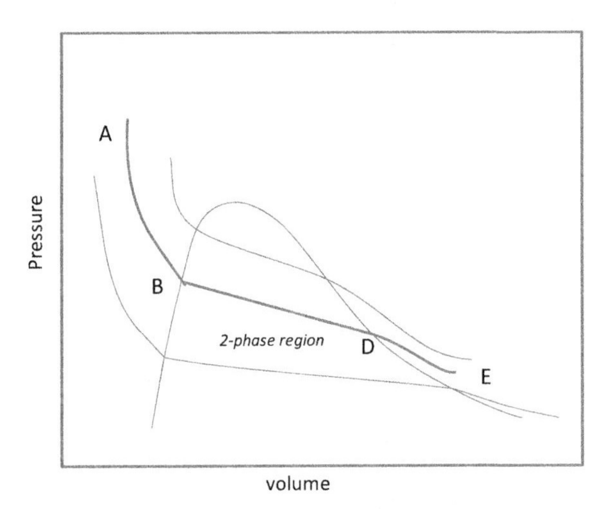

Figure 2.34 PVT in a multicomponent system.

2.7.5.1 Surface Flow Testing

As a well flows to the surface, gas and oil at surface conditions can be recombined in the laboratory to determine all reservoir fluid composition. Production rates are monitored to ensure stable flow rates that can be used in laboratory recombination.

Reservoir fluids should be sampled as early as possible, and sampling should be when the reservoir around the well is above saturation pressure as far as possible—otherwise we never really know the original reservoir composition. This is a particular problem with gas condensates. We try to avoid too high a drawdown, but the rate should be enough to ensure liquid is not collecting in the wellbore. The laboratory recombination setup is shown in Fig. 2.35.

2.7.5.2 Direct Reservoir Fluid Sampling—Repeat Formation Testing

Repeat formation testing (RFT) equipment (Fig. 2.36) is lowered down the wellbore and plugs into the formation taking fluid samples directly (under reservoir conditions). The reservoir fluid sample then goes to the laboratory for testing. This avoids difficulties in ensuring that the correct recombination is used with surface flow testing. The limitation here is the small volumes extracted with RFT sampling, and contamination from drilling fluids.

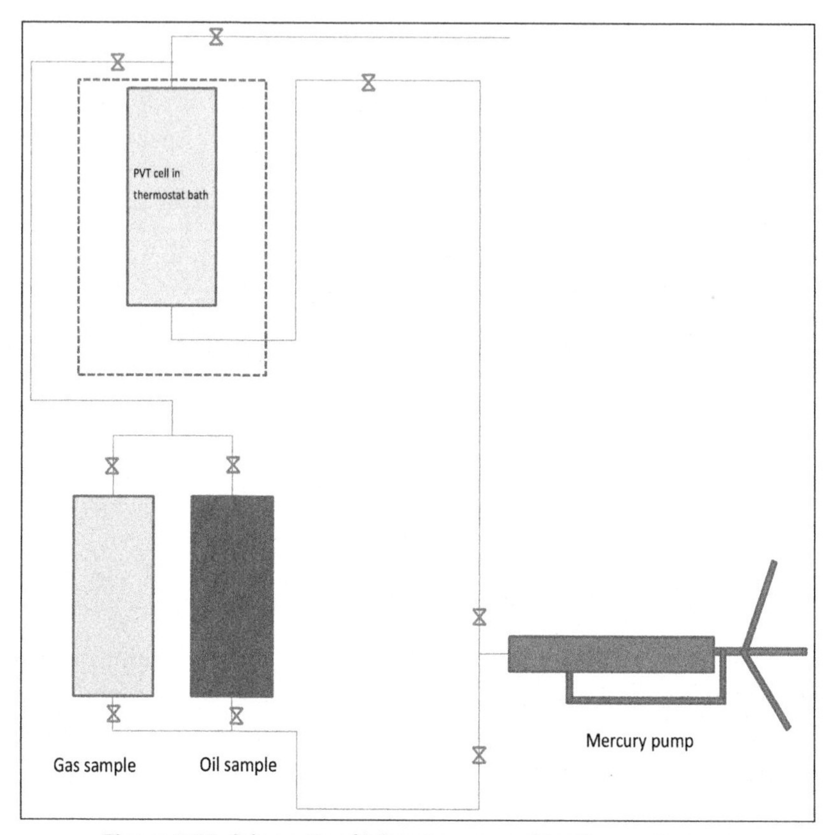

Figure 2.35 Schematic of laboratory recombination equipment.

2.7.6 Laboratory Studies on Reservoir Fluids

For wet gas, dry gas, or gas condensates there are two laboratory experimental methods: constant volume depletion and constant composition expansion (Fig. 2.37).

2.7.6.1 Constant Volume Depletion for Gas and Gas Condensate Systems

A sample of reservoir fluid is maintained in a cell at reservoir temperature, pressure is reduced in stages and gas is removed, bringing the volume back to the original at each stage. Volumes of gas and liquid are measured at each stage. These measurements can then be used to match equations of state to the data for use in compositional modeling, or more directly for black oil tables.

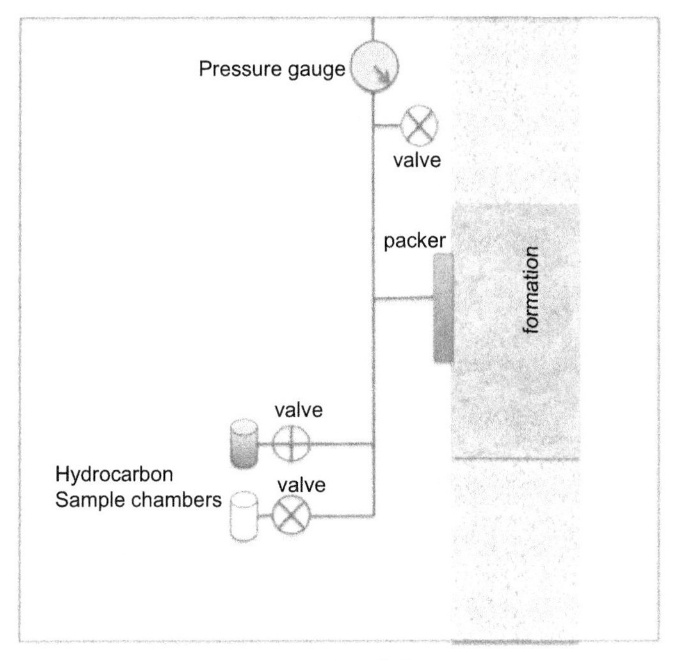

Figure 2.36 Schematic of RFT equipment.

2.7.6.2 Constant Composition Expansion

Here a cell is expanded in a visual cell—pressure is lowered stepwise, with gas and condensate volumes measured at each step. Total composition remains unchanged. These measurements can then be used to match equations of state to the data.

2.7.6.3 Differential Depletion for Oil

Pressure is reduced in stages by expanding a PVT cell volume, with all gas expelled by reducing cell volume to the original volume at constant pressure (Fig. 2.38). This process is continued in stages down to atmospheric pressure. At each stage remaining oil and expelled gas volumes are measured.

Gas formation volume factor (FVF) is calculated from:

$$B_g = V_g / V_{STP} \qquad [2.33]$$

where V_g = gas volume at operating conditions and V_{STP} = volume at standard conditions. Compressibility $Z = (V, P, T_{STP})/(V_{STP}, P_{STP}, T)$, where V = expelled gas volume at test pressure P and temperature T; oil

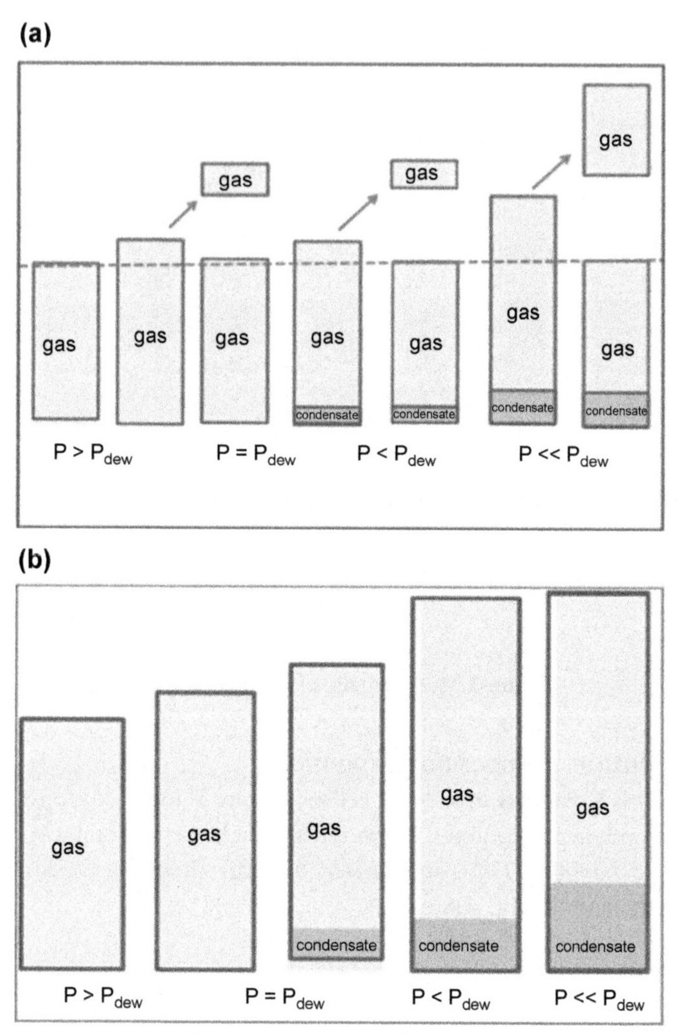

Figure 2.37 (a) Constant volume depletion and (b) constant composition expansion.

formation volume factor B_o = oil volume at operating conditions/volume at standard conditions; and solution GOR (R_s) = total gas evolved at STP/oil volume at STP.

2.7.7 Use of Equations of State in Reservoir Engineering

The *ideal gas law* works well for high temperatures and moderate pressures, and particularly for small molecules (nitrogen, hydrogen, methane) and where molecular attractive forces are small.

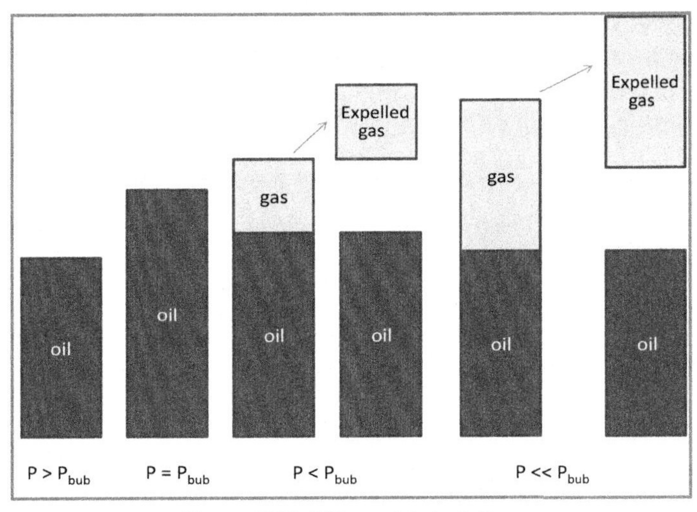

Figure 2.38 Differential depletion.

The law can be theoretically derived if we assume that gas molecules are so small that their actual volume is negligible when compared with total gas volume, and there are no attractive forces between molecules (see Fig. 2.39).

Molecules collide with each other and walls of container and exchange momentum from first principles we can derive the equation:

$$PV = n\mathrm{R}T \qquad [2.34]$$

where P = pressure, V = volume, n = number of moles, T = temperature, and R = gas constant.

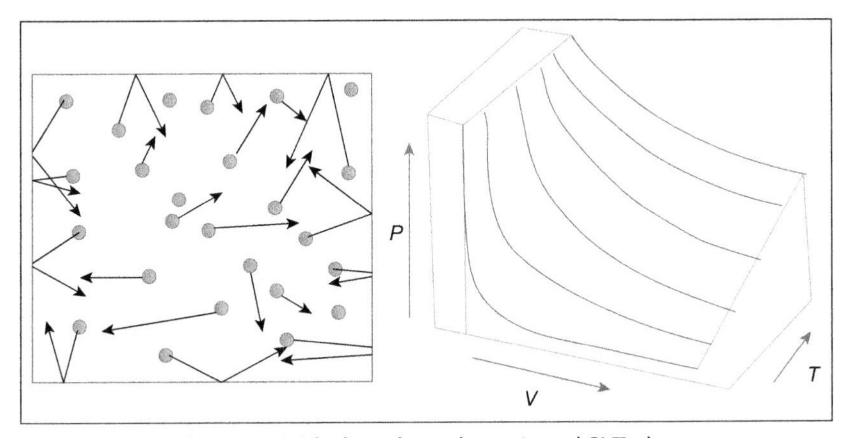

Figure 2.39 Ideal gas law schematic and PVT plot.

2.7.7.1 Real Gases

Allowing for finite size of molecules and attractive intermolecular forces, a number of semi-empirical equations have been derived. An example is the Van der Waals equation:

$$(P + a/V^2)(V - b) = RT \quad [2.35]$$

where constants a and b are dependent on interactive forces and finite molecular volumes. Expanded out in terms of volume V, this gives

$$V^3 - (b + \mathrm{R}T/P)V^2 + (a/P)V - ab/P = 0 \quad [2.36]$$

Or in terms of Z (the deviation factor)

$$Z^3 - (1 + B)Z^2 + AZ - AB = 0 \quad [2.37]$$

and

$$PV = nZ\mathrm{R}T$$

where $A = aP/(\mathrm{R}T)^2$ and $B = bP/\mathrm{R}T$.

These are cubic equations (three solutions at given PT and highest and lowest volumes used).

Fig. 2.40 shows the PVT plot and a Z-factor plot for a typical natural gas. A two-phase region results, and the equation is applicable to liquids as well as gases. To the left of the two-phase region (the all-liquid side) there is only a small change in volume with pressure (liquids have low compressibility), while on the gas side volume is strongly dependent on pressure.

2.7.8 Black Oil Model

For many purposes a reservoir fluid may be represented a simple two-component system of surface gas and surface oil "mixed" in various proportions depending on temperature and pressure (Fig. 2.41).

This is similar to the two-pseudocomponent system shown in Fig. 2.31 above. The difference is that temperature as well as pressure varies on the vertical axis, so the bottom of the schematic represents surface conditions (60°F and 14.7 psi).

Black oil volumetric properties are represented with formation volume factors (FVFs) for gas and oil, while the extent of solution of gas in oil is represented by a solution gas—oil ratio.

2.7.8.1 Formation Volume Factors

Oil and gas expansion factors relating to surface volumes are defined in black oil models.

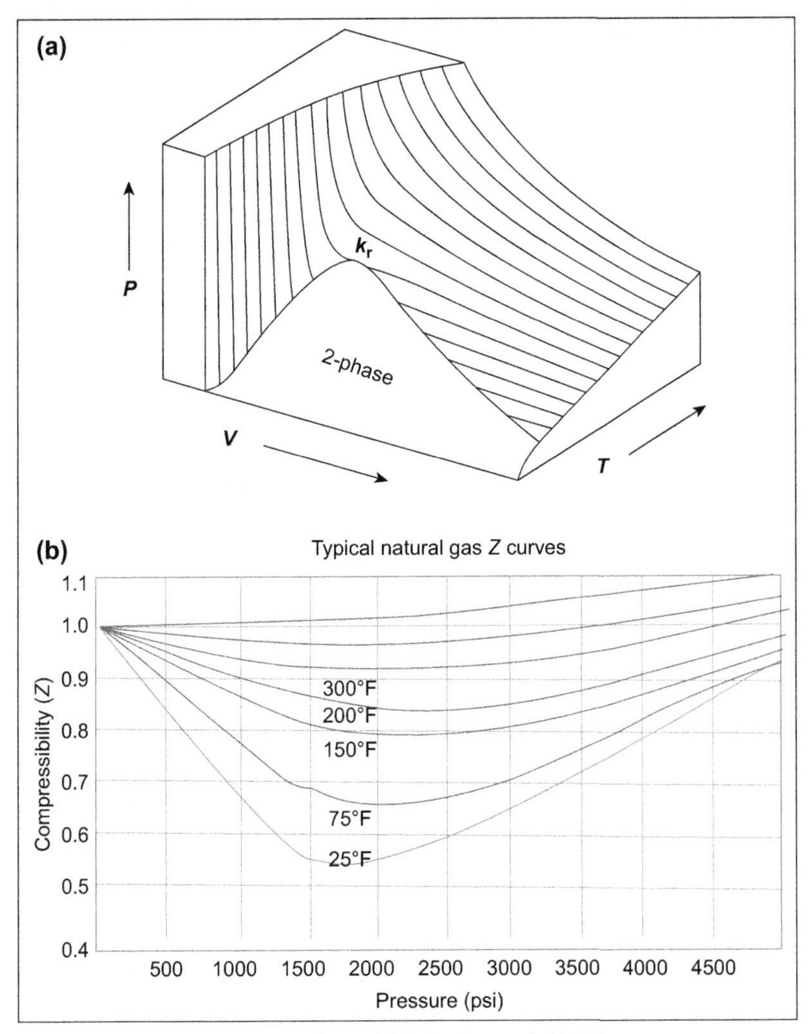

Figure 2.40 Real gas (a) PVT plot and (b) *Z* factor.

2.7.8.1.1 Oil FVF

Oil expands slightly as pressure decreases above the bubble point, and then shrinks as gas is evolved below the bubble point. If we define the oil FVF as

$$B_o(P) = V_o(P)/V_o(\text{STP}) \qquad [2.38]$$

in reservoir barrels per stock tank barrel, where $V_o(\text{STP})$ = volume of oil at standard temperature and pressure, we have a plot like that shown in Fig. 2.42(a), where B_{oi} is the FVF under initial conditions. Note the slight increase in B_o above the bubble point and the faster decrease below the bubble point as pressure decreases.

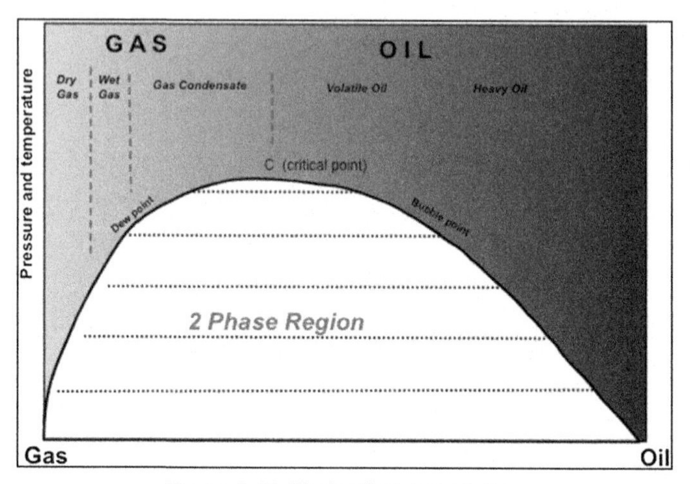

Figure 2.41 Black oil representation.

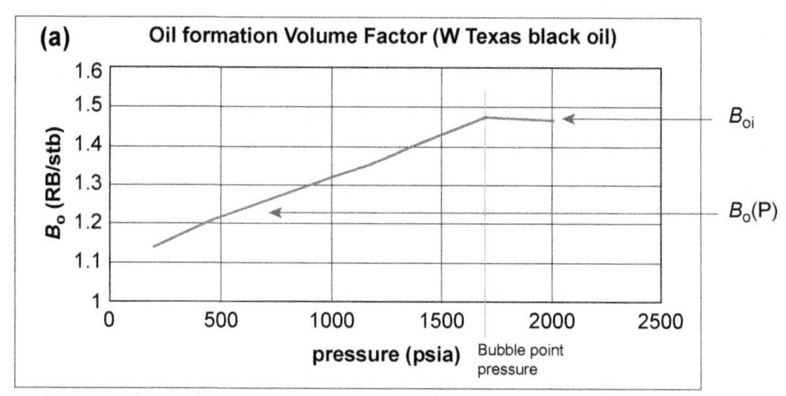

Above bubble point - some v small expansion as pressure drops
Below bubble point - gas released so that oil contracts

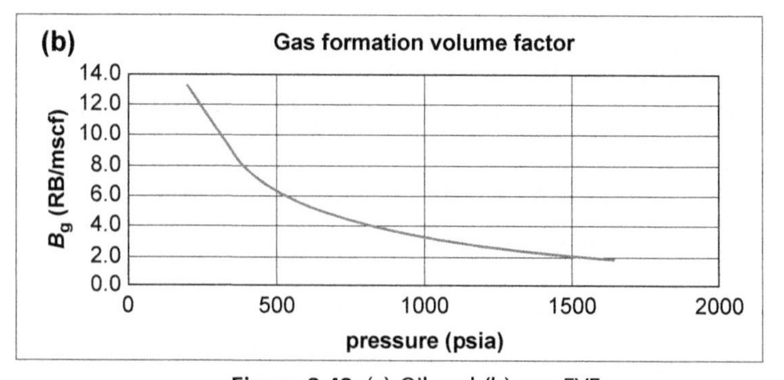

Figure 2.42 (a) Oil and (b) gas FVFs.

2.7.8.1.2 Gas FVF

We have seen above that gas expands as pressure decreases, and we define the gas formation volume factor (B_g) as

$$B_g(P) = V_g(P)/V_g(\text{STP}) \quad [2.39]$$

In field units it can be shown that

$$B_g(p) = 0.0283\ TZ/P$$

2.7.8.2 Solution GOR

We can define a solution GOR as the reciprocal of the gas content of a barrel of oil at a pressure P so that

$$R_s(P) = V_o(P)/V_g(\text{STP}) \quad [2.40]$$

in reservoir barrels/1000 standard cubic feet of released gas.

We get a plot like that shown in Fig. 2.43, where R_s is constant until pressure decreases to the bubble point. Below this pressure the oil can hold less and less gas, so R_s decreases as the volume of gas that the oil can hold decreases.

2.7.9 Excel Software for Producing Empirical Black Oil Curves

A spreadsheet for "black oil properties" is available where empirical equations are used to estimate formation volume factors, solution GORs and viscosities. An example of its use is shown in Fig. 2.44 (input parameters in green cells). It must be understood that for oil these are only very approximate "typical oil" values. For gas we are assuming ideal gas properties. Input for black oil simulators is shown on the right-hand table.

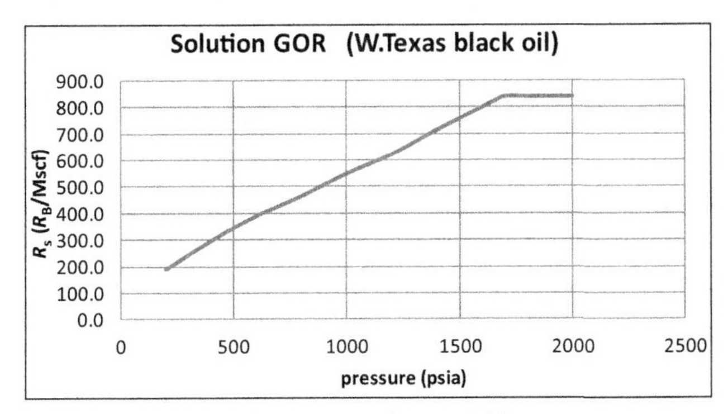

Figure 2.43 Solution GOR.

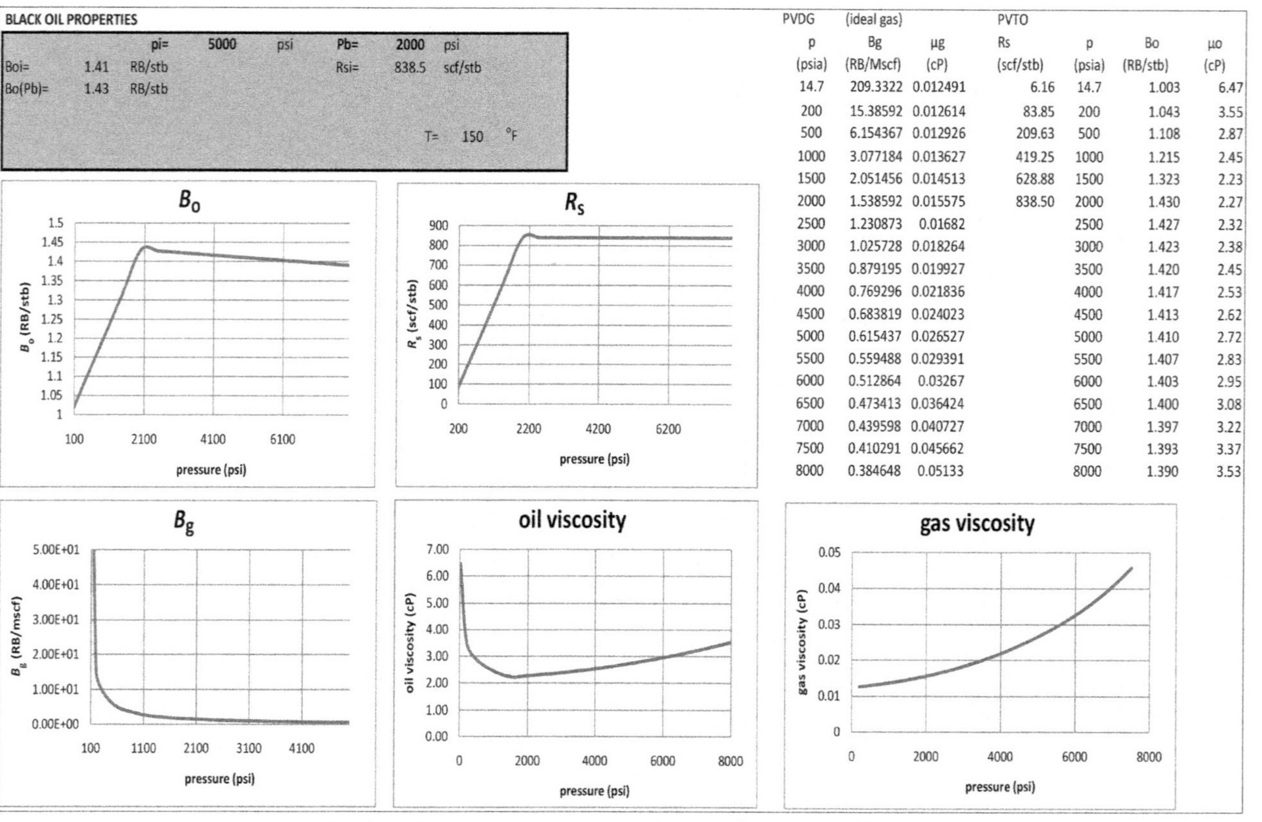

BLACK OIL PROPERTIES

	pi=	5000	psi	Pb=	2000	psi
Boi=	1.41	RB/stb		Rsi=	838.5	scf/stb
Bo(Pb)=	1.43	RB/stb				
				T=	150	°F

PVDG	(ideal gas)		PVTO			
p (psia)	Bg (RB/Mscf)	μg (cP)	Rs (scf/stb)	p (psia)	Bo (RB/stb)	μo (cP)
14.7	209.3322	0.012491	6.16	14.7	1.003	6.47
200	15.38592	0.012614	83.85	200	1.043	3.55
500	6.154367	0.012926	209.63	500	1.108	2.87
1000	3.077184	0.013627	419.25	1000	1.215	2.45
1500	2.051456	0.014513	628.88	1500	1.323	2.23
2000	1.538592	0.015575	838.50	2000	1.430	2.27
2500	1.230873	0.01682		2500	1.427	2.32
3000	1.025728	0.018264		3000	1.423	2.38
3500	0.879195	0.019927		3500	1.420	2.45
4000	0.769296	0.021836		4000	1.417	2.53
4500	0.683819	0.024023		4500	1.413	2.62
5000	0.615437	0.026527		5000	1.410	2.72
5500	0.559488	0.029391		5500	1.407	2.83
6000	0.512864	0.03267		6000	1.403	2.95
6500	0.473413	0.036424		6500	1.400	3.08
7000	0.439598	0.040727		7000	1.397	3.22
7500	0.410291	0.045662		7500	1.393	3.37
8000	0.384648	0.05133		8000	1.390	3.53

Figure 2.44 Example of "black oil properties" spreadsheet.

2.7.10 Compositional Flash Calculations

A "flash" calculation is where we take a known reservoir mixture at known temperature and pressure, and determine the amounts of each phase present, their compositions, and volumetric behavior at some other pressure and temperature.

Using an equation of state such as that of Peng–Robinson, we can take a compositional mixture $\{z_i\}$, where $z_i =$ the mole fraction of component i $\left(\sum_i^N z_i = 1\right)$ and $N =$ the number of components present at a given pressure and temperature, and determine the fractions of liquid and vapor (L and V) and the molar compositions of each of these $\{x_i\}$ and $\{y_i\}$ from basic thermodynamic relationships (see a two-component example in Fig. 2.31).

2.7.10.1 Chemical Potentials

Chemical potentials are the Gibbs free energy per mole for a component I, so that

$$\mu_i = \left(\frac{\partial G}{\partial n_i}\right)_{T,P,nj} \qquad [2.41]$$

and for a system at equilibrium between coexisting liquid and a vapor phase we must have

$$\mu_i^{\mathrm{L}} = \mu_i^{\mathrm{V}} \text{ for all components } i \qquad [2.42]$$

2.7.10.2 Fugacities

From the basic thermodynamic relationship $dG = -SdT + VdP$ and the ideal gas law $PV = \mathrm{R}T$, we can show that for an ideal gas:

$$\left(\frac{\partial \mu}{\partial P}\right)_T = \mathrm{R}T/P$$

so that

$$\mu_i - \mu_i^{\mathrm{o}} = \mathrm{R}T \ln\left(\frac{P}{P_{\mathrm{o}}}\right) \qquad [2.43]$$

where μ_i^{o} is at a reference pressure P_{o}.

2.7.10.3 For a Real Gas

We can define a "corrected pressure" function f_i (for a real gas), called fugacity:

$$\mu_i - \mu_i^o = RT \ln\left(\frac{f_i}{f_i^o}\right) \qquad [2.44]$$

A definition of fugacity is thus as a measure of nonideality. Since $\mu_i^L = \mu_i^V$, it can be shown that similarly, $f_i^L = f_i^V$ for all components i in equilibrium.

We now define a fugacity coefficient φ_i,

$\varphi_i = f_i / P \cdot z_i$ (where z_i = mole fraction of component i)

$\varphi_i \rightarrow 1$ when $P \rightarrow 0$.

From fundamental thermodynamic relationships,

$$\ln(\varphi_i) = \frac{1}{RT} \int \left\{ \left(\frac{\partial P}{\partial n_i}\right)_{T,V,nj} - RT/V \right\} dV - \ln Z \qquad [2.45]$$

Thus once we have a function for Z, we can derive fugacity for a component i in liquid and vapor phases from Z^L and Z^V which come from the equation of state.

Since

$$f_i^V = y_i P \varphi_i^V \text{ and } f_i^L = x_i P \varphi_i^L$$

if we define a ratio of mole fraction of i in the liquid phase to that in the vapor phase:

$$K_i = y_i / x_i = \varphi_i^L / \varphi_i^V$$

Now we also have the relationships $L + V = 1$ and $z_i = x_i \cdot L + y_i \cdot V$.

Using all of the above, the iterative "flash" calculation process can be summarized as detailed in the following subsections.

2.7.10.4 Cubic Equation of State of Form

$$aZ^3 + bZ^2 + cZ + d = 0$$

where a, b, c, and d are functions of the critical properties *of all components* and interaction coefficients between components.

Solved to Give PVT Relationships

This can also give chemical potentials (and fugacities) of each component in gas and liquid phases (μ_i^L and μ_i^V). These must be equal for each component.

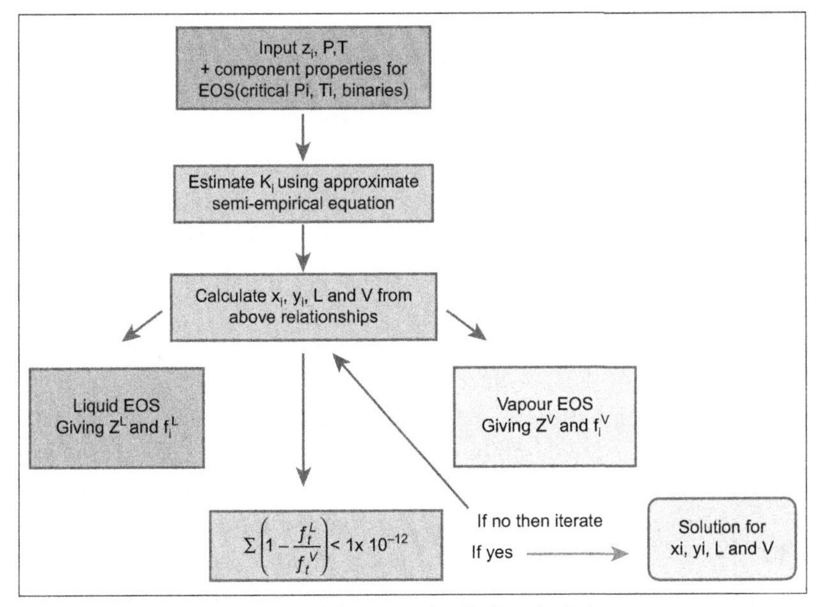

Figure 2.45 Schematic for flash calculations.

2.7.10.5 Allowing Composition of Coexisting Phases to Be Determined

A schematic for standard flash calculations is shown in Fig. 2.45. This type of calculation is used for each grid cell in a compositional reservoir simulator.

2.8 QUESTIONS AND EXERCISES

Q2.1. Define porosity and interconnected pore space. What might be a typical range of values, and what will this depend on?

Q2.2. List two methods by which the porosity at a well is determined.

Q2.3. Define hydrocarbon pore volume.

Q2.4. Define rock compressibility.

Q2.5. Define NTG.

Q2.6. Explain what is meant by the "pore-throat" model of porous rock.

Q2.7. Explain the use of Boyle's law in determining porosity in the laboratory.

Q2.8. Write down Darcy's law, including both pressure gradient and gravitational components. With the aid of a sketch, explain all terms.

Q2.9. Calculate the absolute permeability from laboratory data for an incompressible fluid where $A = 12.5\ cm^2$, $x = 10$ cm, $P_1 - P_2 = 50$ psi, $Q = 0.05\ cm^3/s$, fluid viscosity = 2.0 cP.

Q2.10. Calculate the absolute permeability from the following laboratory data for flow of gas across a horizontal core: $A = 5.06\ cm^2$, $x = 8$ cm, $P_1 = 200$ psi, $P_2 = 195$ psi, $Q = 23.6\ cm^3/s$, gas viscosity = 0.0178 cP.

Q2.11. Write down Darcy's law in field units, listing the dimensions of each.

Q2.12. For an oil—water system, explain with the aid of a sketch the significance of the "contact angle."

Q2.13. Explain with the aid of diagrams what is meant by the terms "drainage" and "imbibition" for an oil—water system (in water wet material). Sketch both capillary pressure and relative permeability curves. To which process would water flooding correspond?

Q2.14. Draw a plot relating oil and water pressures to depth (oil—water system with no gas). Relate this to water saturation.

Q2.15. Explain what is meant by effective permeability. Write the Darcy equation for a phase α where more than a single phase is present.

Q2.16. With the aid of sketches, show the difference you would expect in the relative permeability curves for "water wet" and "oil wet" systems.

Q2.17. Sketch PT phase envelopes for the following.

1. A dry gas.
2. A wet gas.
3. A gas condensate.
4. A heavy oil.

Show changing reservoir conditions and changes to surface conditions.

Q2.18. Describe, with the aid of simple schematics, the laboratory tests for constant volume depletion and constant composition expansion.

Q2.19. The table below shows laboratory measurements of formation volume factors, solution GOR, and viscosities as functions of pressure. Use a spreadsheet to plot all of these. Identify the bubble-point pressure for this oil.

p (psi)	B_o (RB/stb)	B_g (RB/Mscf)	R_s (scf/stb)	μ_o (cp)	M_g (cp)
2000	1.467		838.5	0.3201	
1800	1.472		838.5	0.3114	
1700	1.475		838.5	0.3071	
1640	1.463	1.920	816.1	0.3123	0.016
1600	1.453	1.977	798.4	0.3169	0.016
1400	1.408	2.308	713.4	0.3407	0.014
1200	1.359	2.730	621.0	0.3714	0.014
1000	1.322	3.328	548.0	0.3973	0.013
800	1.278	4.163	464.0	0.4329	0.013
600	1.237	5.471	389.9	0.4712	0.012
400	1.194	7.786	297.4	0.5189	0.012
200	1.141	13.331	190.9	0.5893	0.011

Q2.20. Explain the concept of wettability with the aid of sketches. Write an equation relating contact angle to interfacial energies (explain each term). How is wettability related to the contact angle? What reservoir properties depend on wettability?

Q2.21. Explain, with the aid of a single solid/oil/water interface diagram, the concept of capillary pressure.

2.9 FURTHER READING

L.P. Dake, Fundamentals of Reservoir Engineering, Elsevier, 1978.

L.P. Dake, The Practice of Reservoir Engineering, Elsevier, 2001.

B. Cole Craft, M. Free Hawkins, Applied Petroleum Reservoir Engi neering, Prentice Hall, 2014.

R. Terry, J. Rogers, Applied Petroleum Reservoir Engineering, Prentice Hall, 2015.

A. Kumar, Reservoir Engineering Handbook, SBS Publishers.

P. Donnez, Essentials of Reservoir Engineering, vol. 1 and 2, Editions Technik, 2007 and 2012.

B.F. Towler, Fundamental Principles of Reservoir Engineering, SPE Publications.

W.D. McCain, Properties of Petroleum Fluids, Ebary, 1990.

A. Danesh, PVT and Phase Behavior of Petroleum Reservoir Fluids, Elsevier, 1998.

2.10 SOFTWARE

Relative permeability and capillary pressure.

Black oil properties.

CHAPTER 3

Well-Test Analysis

3.1 INTRODUCTION

Data from log analysis and the core are confined to reservoir properties in the immediate vicinity of the wellbore. Production data and pressure from well testing (transient pressure analysis) enable us to look out much further out into the reservoir.

3.2 BASIC EQUATIONS

We need equations relating pressure (p) at some point at distance (r) from the well (see Fig. 2.7A) to flow rate and time as a function of permeability, porosity, fluid compressibility, and also any boundaries if present.

The starting points are the basic equations of conservation of mass and conservation of momentum.

In terms of radial cylindrical coordinates these are

$$\left(\frac{1}{r}\right)\cdot\frac{\partial}{\partial r}(r\rho u) = \varphi\frac{\partial \rho}{\partial t} \quad [3.1]$$

Conservation of mass and Darcy's law (a semi-empirical law but which can be derived from conservation of momentum with various simplifying assumptions) gives

$$u = -\frac{k}{\mu}\cdot\frac{\partial p}{\partial r} \quad [3.2]$$

where $u =$ velocity, $\varphi =$ porosity, $k =$ permeability, $\rho =$ density, and $\mu =$ viscosity.

A combination of these two equations gives

$$-\frac{1}{r}\cdot\frac{\partial}{\partial r}\left(r\rho\frac{k}{\mu}\frac{\partial p}{\partial r}\right) = \varphi\frac{\partial \rho}{\partial t} \quad [3.3]$$

We now make a set of simplifying assumptions.

Fundamentals of Applied Reservoir Engineering
ISBN 978-0-08-101019-8
http://dx.doi.org/10.1016/B978-0-08-101019-8.00003-X
© 2016 Elsevier Ltd.
All rights reserved.

For oil we assume constant compressibility (c), that permeability, and viscosity are independent of pressure and also that $\frac{\partial p}{\partial r}$ will be small so that $\left(\frac{\partial p}{\partial r}\right)^2$ can be ignored.

This eventually gives us the so-called "diffusivity equation," the basis of all analytical well-test analysis:

$$\frac{\partial^2 p}{\partial r^2} + \frac{1}{r}\frac{\partial p}{\partial r} = \frac{\varphi\mu c}{k}\frac{\partial p}{\partial t} \quad [3.4]$$

which is valid for liquids but not gases. All of the following refers to oil reservoirs. Well-test analysis for gas reservoirs is discussed in appendix "Gas Well Testing."

With a complicated series of mathematical manipulations, this diffusivity equation can be solved under various "boundary conditions" (reservoir conditions).

3.3 LINE SOURCE—INFINITE RESERVOIR

At a point (r, t) in the reservoir (see Fig. 3.1):

$$p(r,t) = p_{\mathrm{i}} - \frac{qB\mu}{2\Pi kh}\cdot\frac{1}{2}\left\{\ln\left(\frac{kt}{\varphi\mu cr^2}\right) + 0.80907\right\} \quad [3.5]$$

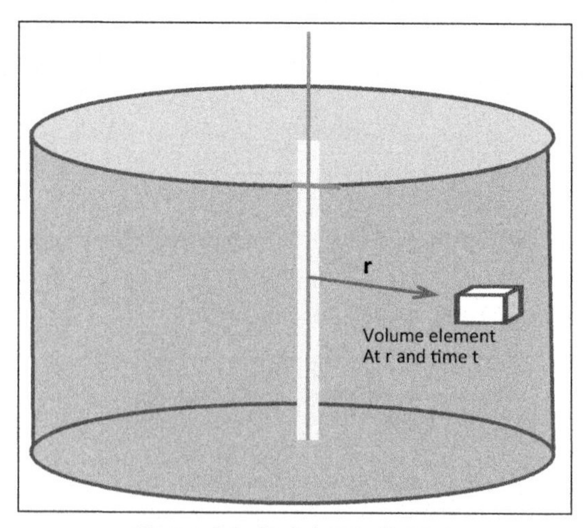

Figure 3.1 Radial coordinates.

and at the well ($r = r_w$):

$$p_w(t) = p_i - \frac{qB\mu}{2\Pi kh} \cdot \frac{1}{2} \left\{ \ln\left(\frac{kt}{\varphi \mu c r_w^2}\right) + 0.80907 \right\} \quad [3.6]$$

where $p_w(t)$ = pressure at the wellbore at time t; p_i = the initial pressure; B = oil formation volume factor; and r_w = wellbore radius.

In field units this equation is:

$$p_w(t) = p_i - \frac{162.6 qB_o\mu}{kh} \cdot \left\{ \log_{10}\left(\frac{kt}{\varphi \mu c r_w^2}\right) - 3.23 \right\} \quad [3.7]$$

where p_w and p_i are in psi; q in stb/d; μ in cP; k in mD; h in ft; $c = c_w S_w + c_o S_o + c_f$ in psi^{-1}; r in ft; B_o in RB/stb; and t in hours.

3.4 BOUNDED RESERVOIR WITH "NO FLOW" BOUNDARY

That is a closed system.

Solution of the diffusivity equation with appropriate boundary conditions gives

$$p_w(t) = p_i - \frac{qB\mu}{2\Pi kh} \cdot \left\{ \left(\frac{2kt}{\varphi \mu c r_e^2}\right) + \ln\left(\frac{r_e}{r_w}\right) - 0.75 \right\} \quad [3.8]$$

where r_e = the boundary radius.

3.5 CONSTANT PRESSURE BOUNDARY

Again we solve the diffusivity equation with suitable boundary conditions, giving

$$p_w(t) = p_i - \frac{qB\mu}{2\Pi kh} \ln\left(\frac{r_e}{r_w}\right) \quad [3.9]$$

This will correspond to a "steady-state" system independent of time, and rearranged gives the Darcy law equation for a radial cylindrical system discussed in chapter "Basic Rock and Fluid Properties."

3.6 SKIN EFFECTS

Up to now we have considered the pressure (p_w) just adjacent to the wellbore. This will not necessarily be the same as that within the wellbore itself (p_{wf}). The pressures will normally be different, because damage or improvement to flow properties can occur when completing the well. For example, drilling mud can infiltrate the formation, decreasing permeability, so that we have larger than expected pressure drops. Alternatively some local fracturing around the well can increase permeability. These "skin effects" are normally very local to the well (perhaps a few inches or feet), thus affecting only a small volume of the reservoir (Fig. 3.2).

In treating skin effect we assume that

$$p_{wf}(t) = p_w(t) + \Delta p_s \qquad [3.10]$$

If we assume that the pressure drop is proportional to flow rate and that we have steady-state flow in this region, then,

$$S = \Delta p_s / (qB\mu / 2\Pi kh) \qquad [3.11]$$

where k is the local permeability in the altered zone.

For an infinite acting reservoir with the well producing at a constant rate (q) we then have

$$p_{wf}(t) = p_i - \frac{qB\mu}{2\Pi kh} \cdot \frac{1}{2} \left\{ \ln\left(\frac{kt}{\varphi \mu c r_w^2}\right) + 0.80907 + 2S \right\} \qquad [3.12]$$

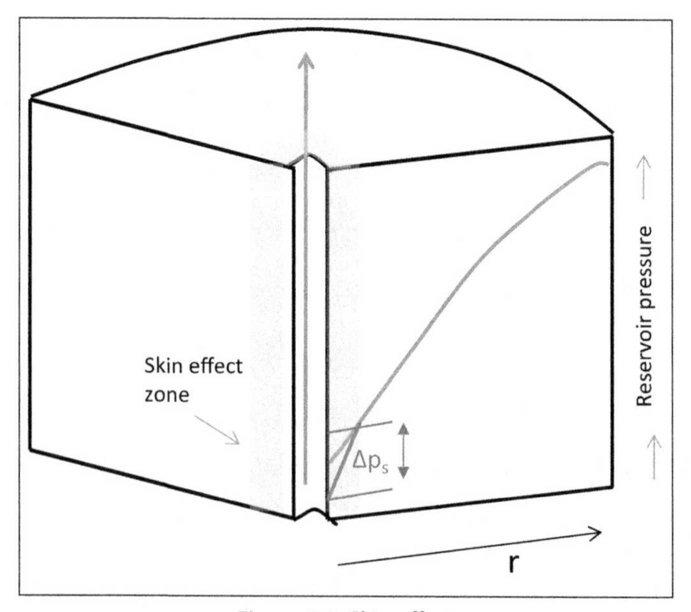

Figure 3.2 Skin effect.

or in field units:

$$p_w(t) = p_i - \frac{162.6\, qB_o\mu}{kh} \cdot \left\{ \log_{10}\left(\frac{kt}{\varphi\mu c r_w^2}\right) - 3.23 + 0.87S \right\} \qquad [3.13]$$

and for a bounded reservoir (radius $= r_e$):

$$p_{wf}(t) = p_i - \frac{qB\mu}{2\Pi kh} \cdot \left\{ \left(\frac{2kt}{\varphi\mu c r_e^2}\right) + \ln\left(\frac{r_e}{r_w}\right) - 0.75 + S \right\} \qquad [3.14]$$

Skin effects are discussed further in Section 3.9.

3.7 WELLBORE STORAGE

A further complicating factor is that the well is normally opened and shut at the surface, while the pressure gauge is at the bottom of the well (Fig. 3.3). Hence when flow is started or stopped there is a period while fluid in the wellbore responds to the opening or shutting of the wellhead valve. The time required for reservoir flow to be established depends on the volume in the wellbore and the compressibility of the fluids (in particular the effect is larger if gas is present).

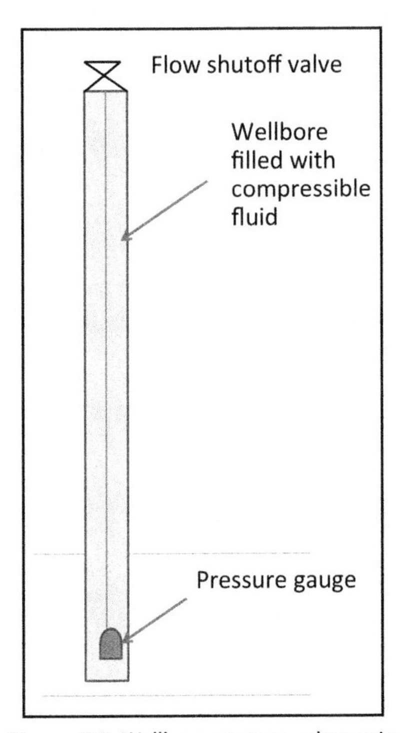

Figure 3.3 Wellbore storage schematic.

Hence shut-off provokes an immediate response within the compressible fluid in the wellbore, as pressure initially increases in the wellbore giving a reduction in volume. Early pressure measured is thus distorted by a wellbore storage effect.

3.8 PRESSURE DRAWDOWN ANALYSIS

From the above

$$p_{\mathrm{wf}}(t) = p_{\mathrm{i}} - \frac{qB\mu}{2\Pi kh}\cdot\frac{1}{2}\left[\ln(t) + \ln\left(\frac{k}{\varphi\mu c r_{\mathrm{w}}^{2}}\right) + 0.80907 + 2S\right] \quad [3.15]$$

or in field units:

$$p_{\mathrm{w}}(t) = p_{\mathrm{i}} - \frac{162.6\, qB_{\mathrm{o}}\mu}{kh}\cdot\left\{\log_{10}\left(\frac{kt}{\varphi\mu c r_{\mathrm{w}}^{2}}\right) - 3.23 + 0.87S\right\} \quad [3.16]$$

Therefore we have the general relationship form for well pressure, initial reservoir pressure, and time t:

$$p_{\mathrm{wf}}(t) = m\cdot\log(t) + p_{\mathrm{i}} \quad [3.17]$$

If we consider a period where a well is shut-in followed by a constant flow rate period, we have the position shown in Fig. 3.4.

There is a slope m in the straight-line portion of the curve p_{wf} versus log(t), where $m = \frac{qB\mu}{4\Pi kh}$, which gives us permeability k, where B (formation volume factor) and viscosity are known from pressure/volume/temperature data and also h and q are known.

Also consider log(t) $= 0$, ie, $t = 1$, from which the skin factor S can in principle be obtained, although wellbore storage complicates this.

The basic form of the solutions to the diffusion equations can be written as

$$\Delta p = p_{\mathrm{wf}} - p_{1} = q\cdot m\cdot f(t) \quad [3.18]$$

3.9 PRESSURE BUILDUP ANALYSIS

3.9.1 The Principle of Superposition

For general use we need to handle situations where we may have more than one well, or more commonly a single well producing at a varying rate. Fortunately, for linear differential equations, the same relationships discussed above can still be used, but with appropriate boundary conditions.

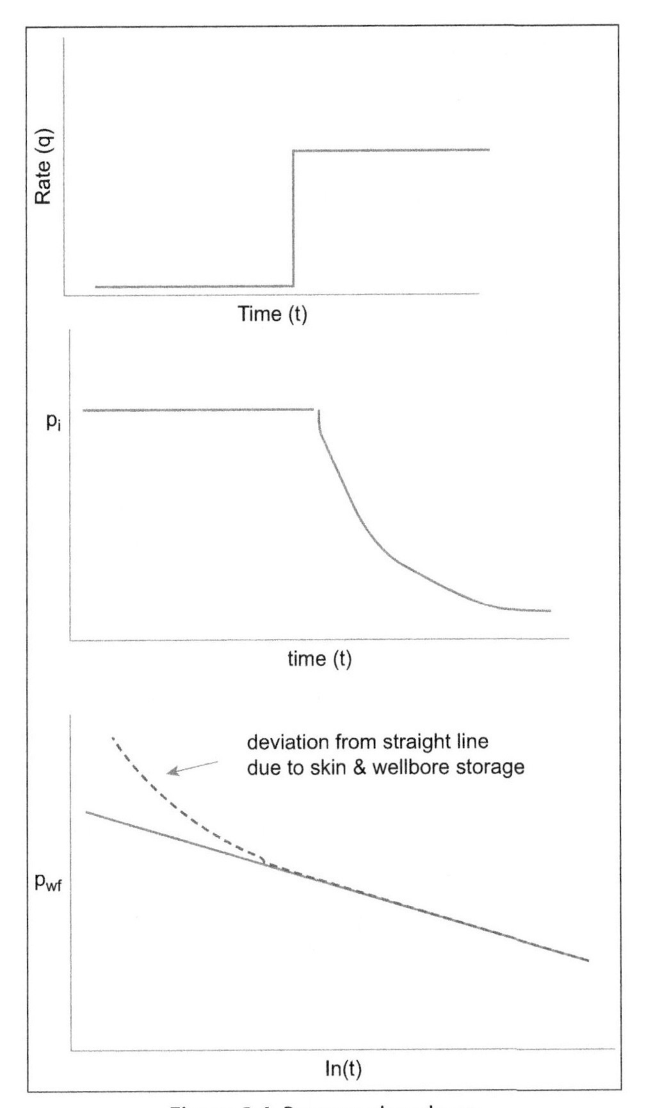

Figure 3.4 Pressure drawdown.

Consider a single well with a drawdown period followed by a shut-in and consequent buildup (Fig. 3.5).

3.9.2 Horner Plots—Permeability and Initial Pressure From Pressure Buildup Data

If Δp = pressure buildup during Δt,

$$\Delta p = m\{q_1 f(t) + (q_2 - q_1) f(t - t_p)\} \tag{3.19}$$

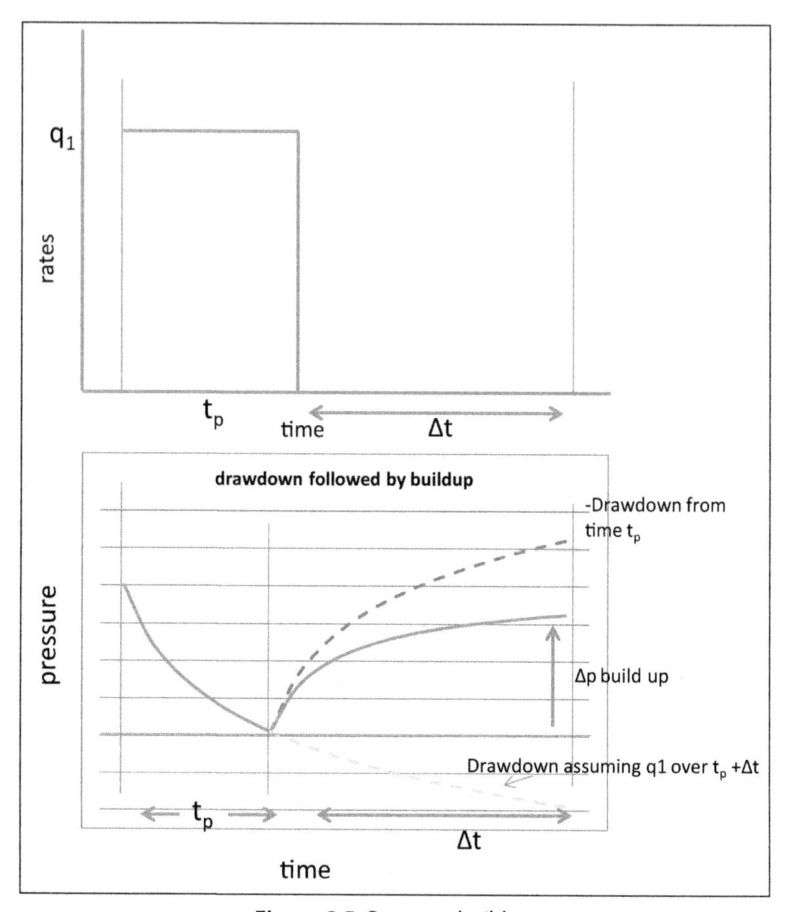

Figure 3.5 Pressure buildup.

if $q_2 = 0$,

$$\Delta p = q_1 \cdot m\{f(t) - f(t - t_p)\} \qquad [3.20]$$

if $t = t_p + \Delta t$,

$$\Delta p = q_1 \cdot m\{f(t_p + \Delta t) - f(\Delta t)\} \qquad [3.21]$$

Since,

$$f(\Delta t) = \frac{1}{2}(\ln(\Delta t) + c) \quad \text{and} \quad f(t) = \frac{1}{2}(\ln(t_p + \Delta t) + c) \qquad [3.22]$$

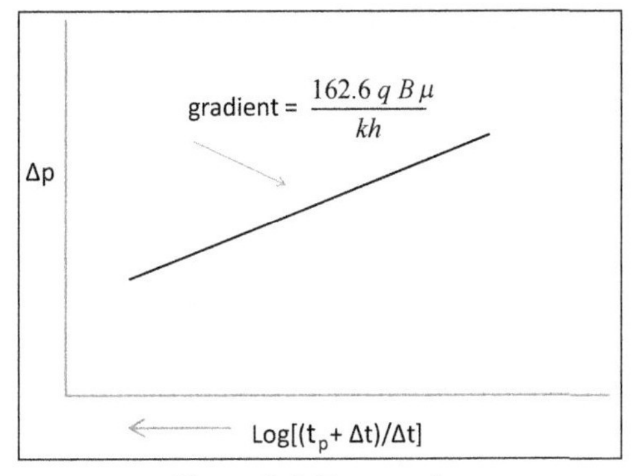

Figure 3.6 Horner plot.

$$\Delta p = q \cdot \frac{m}{2} \cdot \ln\{(t_p + \Delta t)/\Delta t\} \quad [3.23]$$

This is known as the Horner buildup equation (Fig. 3.6).

In field units,

$$\Delta p = \frac{162.6\, qB\mu}{kh} \log\{(t_p + \Delta t)/\Delta t\} \quad [3.24]$$

Eq. [3.24], however, does not allow for early time effects—the skin and wellbore storage factors discussed above. Fig. 3.7 shows a typical pressure buildup plot with early time dominated by wellbore storage and skin effects. The plot also shows the Horner equation matched to the middle time linear section of the pressure data when wellbore storage and skin effects have disappeared. Estimated permeability can be varied to give this match and also extrapolation down to 1 h will give initial reservoir pressure. A spreadsheet discussed in Section 3.12 is available to do this.

3.9.3 Skin Factor From Buildup Data

We assumed in defining skin factor S that we have steady-state flow very close to the wellbore. If this is assumed to show up in the very earliest pressure buildup data and it is conventional to use data at t = 1 h, Eq. [3.16] can be rearranged to give an estimate for S:

$$S = 1.151 \left[\frac{p_{1\text{ hr}} - p_{\text{wf}}}{\frac{kh}{162.3 q B_o \mu}} - \log_{10} \frac{k}{\phi \mu c r_w^2} + 3.23 \right] \quad [3.25]$$

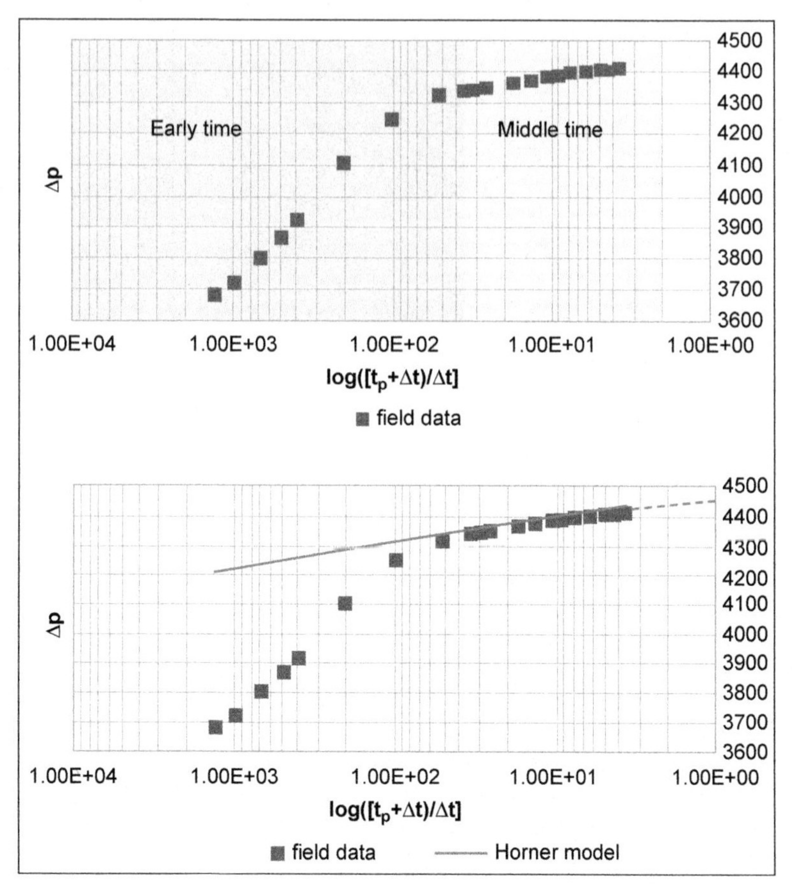

Figure 3.7 Horner plots—use of Horner equation to match field data.

3.10 LOG–LOG PLOTS—MOST COMMONLY USED ANALYSIS TOOL

Where Δt is large, the derivative of this can be approximated as

$$\frac{d[\ln(\Delta p)]}{d[\ln(\Delta t)]} = \frac{q\mu}{4\pi kh} \quad [3.26]$$

to get log–log derivative plots (see the red line (light gray in print versions) in the plot in Fig. 3.8), which are the most common analysis tool currently used. Infinitely acting radial flow (IARF) will thus give a value for $\frac{q\mu}{4\pi kh}$ from which k can be determined. Note that k is in the denominator, so that the higher the permeability the lower the height of the derivative line at radial flow.

Log–log plots of pressure change and its derivative versus buildup time are now the most common tools, along with the Horner plot discussed above, in routine analysis of well-test data. There are a number of

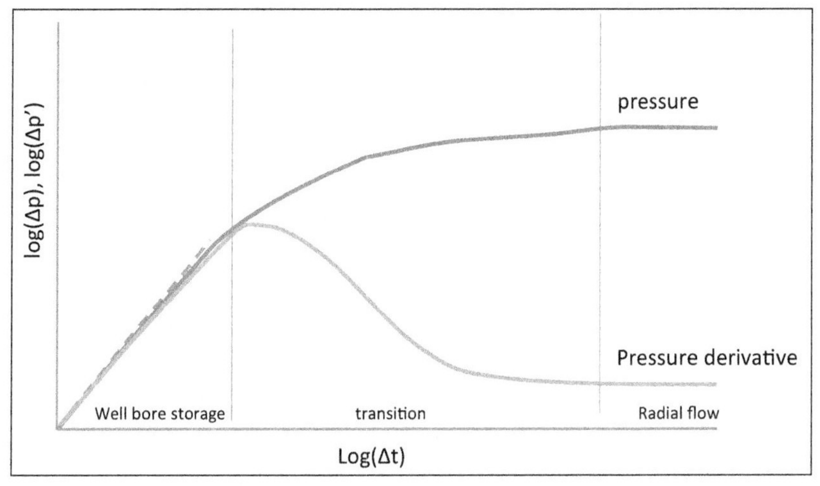

Figure 3.8 Log—log derivative plot.

commercial software packages that use **both analytical and numerical programs** to compare and match possible reservoir models against field data using a range of boundary conditions.

3.11 RESERVOIR TYPES

The measured data are matched to reservoir types, examples of which are discussed below.

3.11.1 Radial Composite Models

The radial composite model identifies variations in transmissibility between the region close to the well, the inner IARF and the IARF further away. Where transmissibility decreases in the outer IARF zone, the derivative line has a higher value than that for the inner zone (see Fig. 3.9). The opposite is the case when permeability increases further from the well. This transmissibility change may be due to a change in permeability, but it can also result from partial fault barriers.

3.11.2 Constant Pressure Boundary

Where we have a constant pressure boundary, for example, where we have pressure support from a powerful aquifer, the derivative line shows a decline as pressure stabilizes (Fig. 3.10).

3.11.3 Closed Radial System

Where we have a closed system (Fig. 3.11) the buildup plots are similar to that for a constant pressure boundary system, and must be distinguished using analysis of drawdown curves.

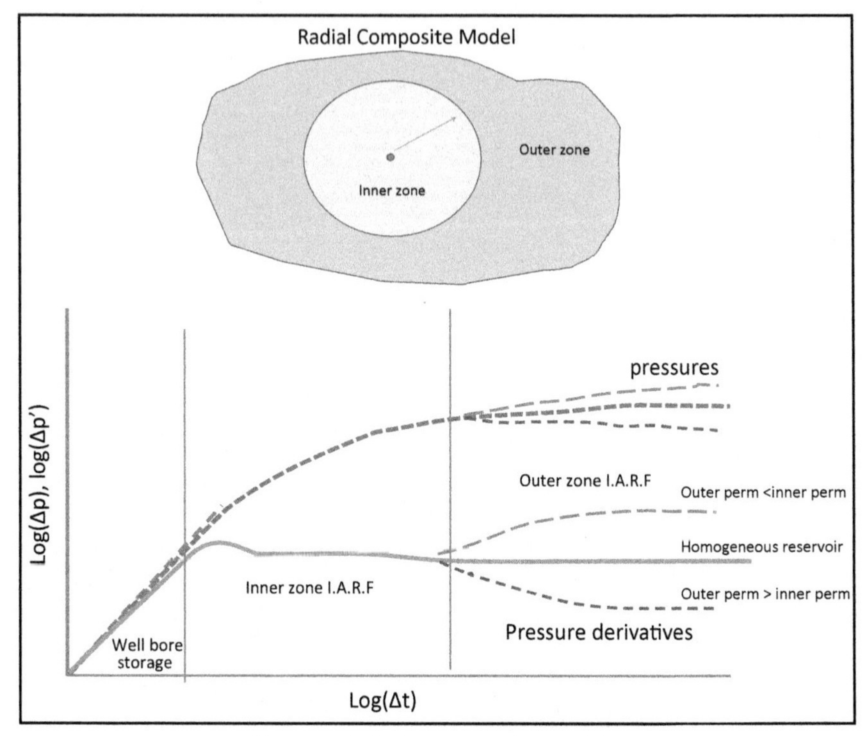

Figure 3.9 Radial composite models.

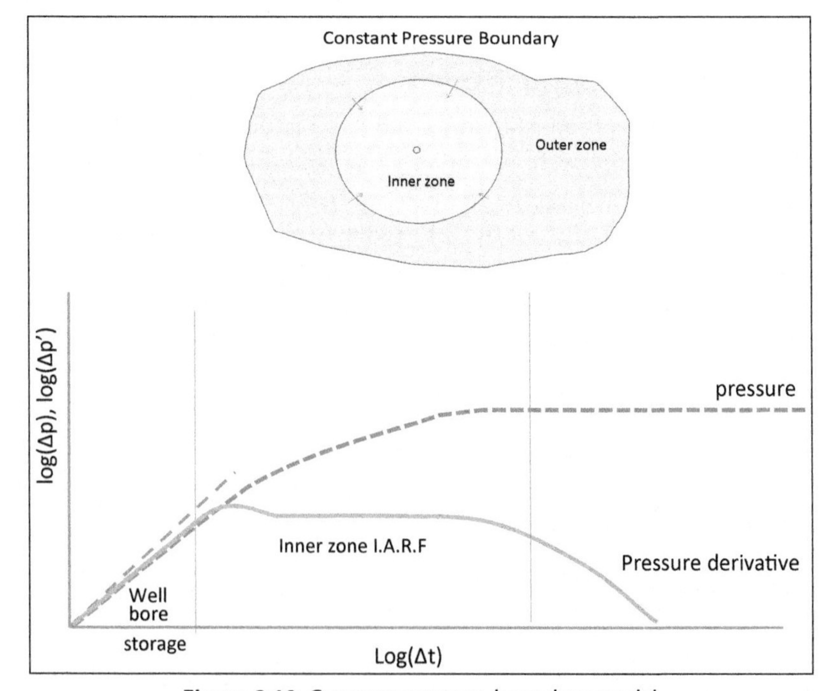

Figure 3.10 Constant pressure boundary model.

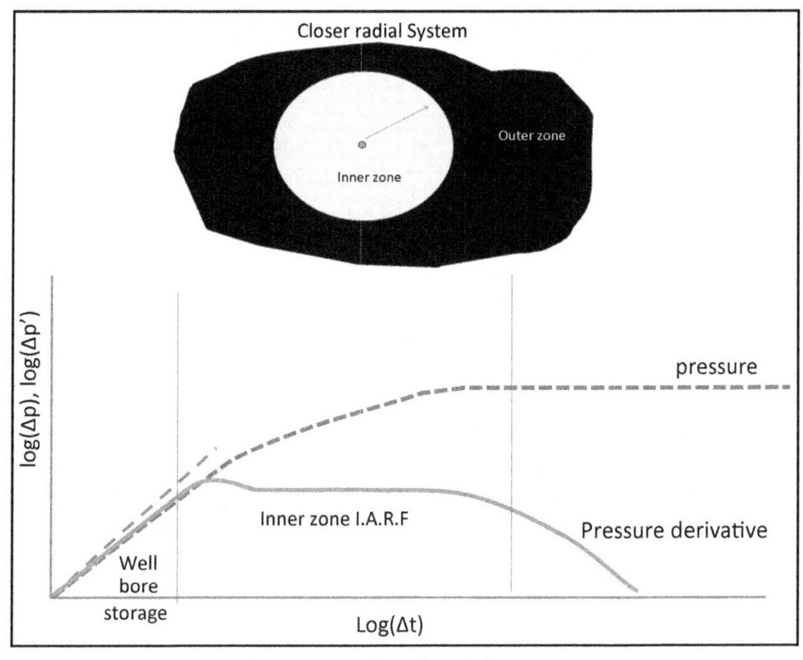

Figure 3.11 Closed radial system.

3.11.4 Fractured Reservoir

A heavily fractured reservoir will have a buildup plot like that shown in Fig. 3.12.

3.12 EXCEL SPREADSHEET FOR PRESSURE BUILDUP ANALYSIS

A useful Excel spreadsheet, "horner plot-zz," generating a Horner plot is available. An example with input and output is shown in Fig. 3.13.

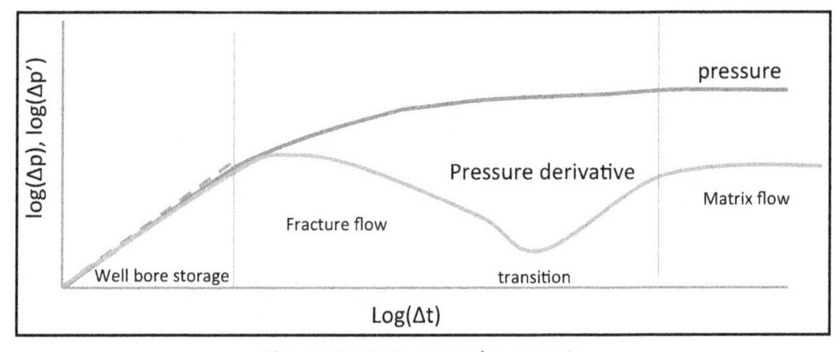

Figure 3.12 Fractured reservoir.

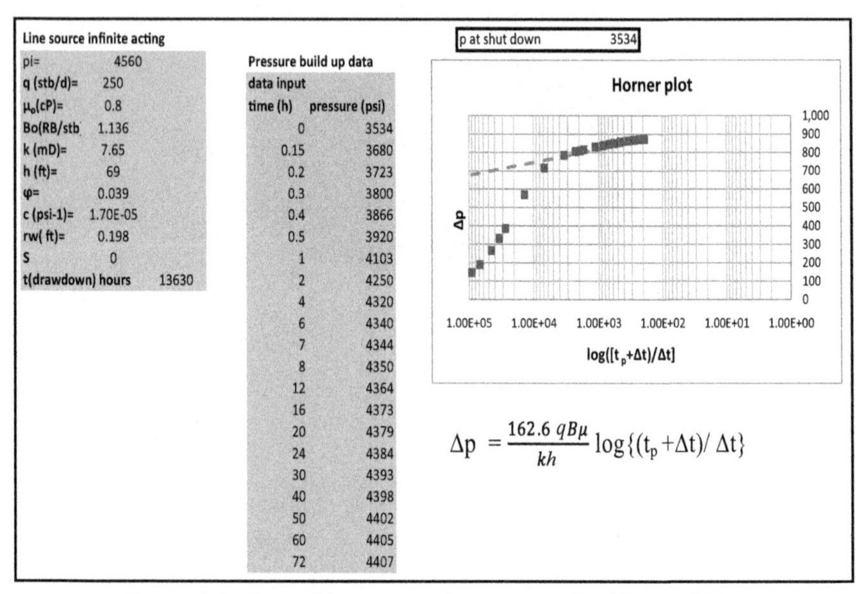

Figure 3.13 Spreadsheet example—pressure buildup analysis.

3.13 QUESTIONS AND EXERCISES

Q3.1. The diffusivity equation in terms of radial coordinates for slightly compressible fluids, which is the equation forming the basis for much well-test analysis, is shown below:

$$\frac{\partial^2 p}{\partial r^2} + \frac{1}{r}\frac{\partial p}{\partial r} = \frac{\varphi \mu c}{k}\frac{\partial p}{\partial t}$$

Explain the meaning of each of the terms in this equation. Name one assumption made in deriving this equation.

Q3.2. Name three boundary conditions commonly used to explain different reservoir types.

Sketch a log–log pressure and derivative plot for a heavily, naturally fractured reservoir.

Q3.3. We have the oil flow rate versus time shown in the table below:

Time (h)	0	1.5	3	6	9	12	18	24	48	72
Pressure (psi)	5050	4943	4937	4935	4929	4927	4923	4921	4916	4912

Oil viscosity = 0.5 cP, oil formation volume factor = 1.75 RB/stb, oil rate = 500 bopd, formation thickness = 60 ft, porosity = 0.2, compressibility = 1.5×10^{-5}, $r_w = 0.16$ ft, and skin = 0.0.

With the software provided (welltest analysis-drawdown-zz), vary the permeability (k) value to match the data and thus obtain the permeability of the formation.

Q3.4. A pressure buildup test was carried out for 100 h with an initial pressure of 4800 psi, with the results shown in the table below.

Time (h)	0	10	20	30	40	50	60	70	80	90	100
Pressure (psi)	4800	4890	4920	4950	4968	4972	4975	4977	4979	4981	4982

Oil viscosity = 0.5 cP, oil formation volume factor = 1.80 RB/stb, oil rate = 400 bopd, formation thickness = 50 ft, porosity = 0.2, compressibility = 1.5×10^{-5}, $r_w = 0.16$ ft, and skin = 0.0. Initial reservoir pressure was 5000 psi.

With the software provided (horner plot-zz), vary the permeability (k) value to match the data and thus obtain the permeability of the formation.

3.14 FURTHER READING

J. Lee, Well testing, SPE Textbooks, 1982.
J. Spivey, J. Lee, Applied Well Test Interpretation, SPE Textbooks, 2013.

3.15 SOFTWARE

Welltest drawdown
Horner plot

CHAPTER 4

Analytical Methods for Prediction of Reservoir Performance

4.1 INTRODUCTION

From fundamental equations such as conservation of mass and momentum (Darcy's law) and thermodynamic relationship we can derive analytical equations, such as those for material balance and the Buckley–Leverett equation for water advance in water flooding that have historically been used by reservoir engineers to estimate oil and gas recovery as a function of pressure decrease. Also simple single-cell well models can be used in early field appraisal and development planning. With the increasing speed and sophistication of numerical simulations, use of these methods has declined, although they are still very useful in understanding the dynamics of reservoir behavior and can be used in conjunction with decline curve methodology in early evaluation of potential recovery factors and development plans.

4.2 DECLINE PERFORMANCE FROM MATERIAL BALANCE

For both gas and oil we can use analytical material balance equations to predict reservoir performance with decreasing pressure. In both cases, to predict production as a function of time (ie, to obtain a production profile), we need to make assumptions on the change in pressure with time.

4.2.1 Material Balance for Gas Reservoirs

4.2.1.1 Gas Equation of State

$$PV = nZRT \qquad [4.1]$$

Since,

$$p_i V_H = n_i Z_i RT \qquad [4.2]$$

at initial conditions, where p_i = initial pressure; V_H = hydrocarbon volume; n_i = number of moles in reservoir at initial conditions; and Z_i = compressibility factor at initial conditions ($Z_i = f(p)$).

Fundamentals of Applied Reservoir Engineering
ISBN 978-0-08-101019-8
http://dx.doi.org/10.1016/B978-0-08-101019-8.00004-1
© 2016 Elsevier Ltd.
All rights reserved.

$$pV_H = nZRT \qquad [4.3]$$

where p = pressure at some time t; n = number of moles in reservoir at time t; and Z = compressibility factor at time t.

If Δn_i = number of mole produced up until time t, then $(p/Z)n_i = (p_i/Z_i)(n_i - \Delta n_i)$ or

$$\frac{p}{Z} = \frac{p_i}{Z_i}\left(1 - \frac{\Delta n_i}{n_i}\right) \qquad [4.4]$$

Now,

$$p^o = n_i Z^o R\, T^o / V_i^o \qquad [4.5]$$

Relating gas volume at standard conditions to a number of moles initially present.

$$p^o = \Delta n_i Z^o R T^o / \Delta V^o \qquad [4.6]$$

Relating volume of gas produced at standard conditions to number of moles produced:

$\frac{\Delta n_i}{n_i} = \frac{\Delta V^o}{V_i^o}$ and therefore,

$$\frac{p}{Z} = \frac{p_i}{Z_i}\left(1 - \frac{\Delta V^o}{V_i^o}\right) \qquad [4.7]$$

giving a simple linear relationship for volume of gas produced when going from initial pressure p_i to a final pressure p

$$\Delta V^o = V_i^o\left(1 - \frac{pZ_i}{Zp_i}\right) \qquad [4.8]$$

where ΔV^o = volume of gas produced (bscf − surface conditions); V_i^o = gas initially in place (bscf − surface conditions); p_i = initial pressure (psi); p = final pressure (psi); Z_i = compressibility at initial conditions ($Z_i = f(p_i)$); and Z = compressibility under final conditions ($Z = f(p)$).

4.2.2 Diagnostics—Determination of Gas Initially in Place

p/Z versus ΔV_o plots can give estimates of total reservoir volumes from early production data. The vertical intersect p_i/Z_i and the horizontal axis intersect give the total gas in place. We thus can use early production data (ΔV_o) and laboratory pressure/volume/temperature data (for Z) to estimate the gas in place (see Fig. 4.1a).

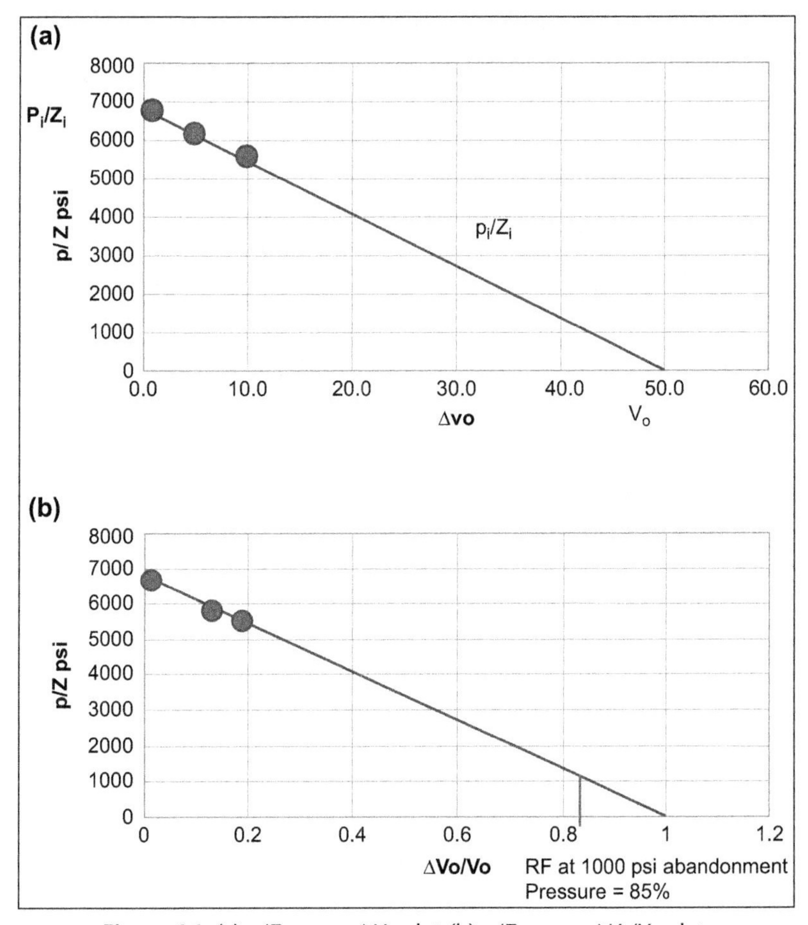

Figure 4.1 (a) p/Z versus ΔV_o plot (b) p/Z versus $\Delta V_o/V_o$ plot.

Recovery factors at various abandonment pressures can be obtained from early p/Z versus $\Delta V_o/V_o$ data (see Fig. 4.1b).

4.2.3 Material Balance for Oil Reservoirs

A volumetric balance at reservoir conditions gives total produced oil.

1. Expansion of original oil between p_i and p is $= N(B_o - B_{oi}) =$ expansion oil drive.
2. Expansion of liberated gas $= NB_g(R_{si} - R_s) =$ solution gas drive.

3. Change in hydrocarbon pore volume between p_i and $p = NB_{oi} \cdot \Delta p(c_w \cdot S_{wi} + c_f)/(1 - S_{wi})$ due to expansion of connate water and the grain arrangement = compaction drive.
4. Any net influx (water and/or gas) = W_i.

To find the stock tank cumulative volume of oil produced between p_i and p (=ΔN), ΔN = expansion of original oil + expansion of liberated gas + change in pore volume + any net influx, and thus,

$$\Delta N = [N \cdot \{(B_o - B_{oi}) + (R_{si} - R_s) \cdot B_g + \Delta p \cdot B_{oi} \cdot (c_w S_{wi} + c_f)/(1 - S_{wi})\} + W_f]/\{B_o + (R_p - R_s) \cdot B_g\} \quad [4.9]$$

where N = stock tank barrels (stb) of oil initially in place; ΔN = stb of oil produced; B_{oi} = initial formation volume factor (fvf) in RB/stb, B_o = fvf at lower pressure (RB/stb); R_{si} = initial solution gas–oil ratio (GOR) (scf/stb, R_s = GOR at some other pressure (scf/stb); R_p = cumulative produced GOR (scf/stb); B_g = gas fvf (RB/scf); c_w = water compressibility; and S_{wi} = initial connate water saturation. This gives the oil produced (in stb) resulting from a pressure drop from p_i to p.

To estimate final recovery we need to assume an abandonment pressure from surface operating conditions and wellbore hydraulics.

4.2.4 Diagnostics—Determination of Oil Initially in Place: Havlena–Odeh Analysis

Rearranging the above basic equation, if

$$F = \Delta N \cdot \{B_o + (R_p - R_s) \cdot B_g\}$$
$$E_o = \{(B_o - B_{oi}) + (R_{si} - R_s) \cdot B_g\}$$
$$E_g = B_{oi} \cdot (B_g/B_{gi} - 1)$$

then,

$$F = N \cdot [E_o + mE_g]$$

and a plot of F versus $[E_o + mE_g]$ should be a straight line with **slope N** = bbl initial.

The vertical axis represents a “withdrawal term” and the horizontal axis an “expansion term.”

In the example shown in Fig. 4.2, the slope = 45 mmstb.

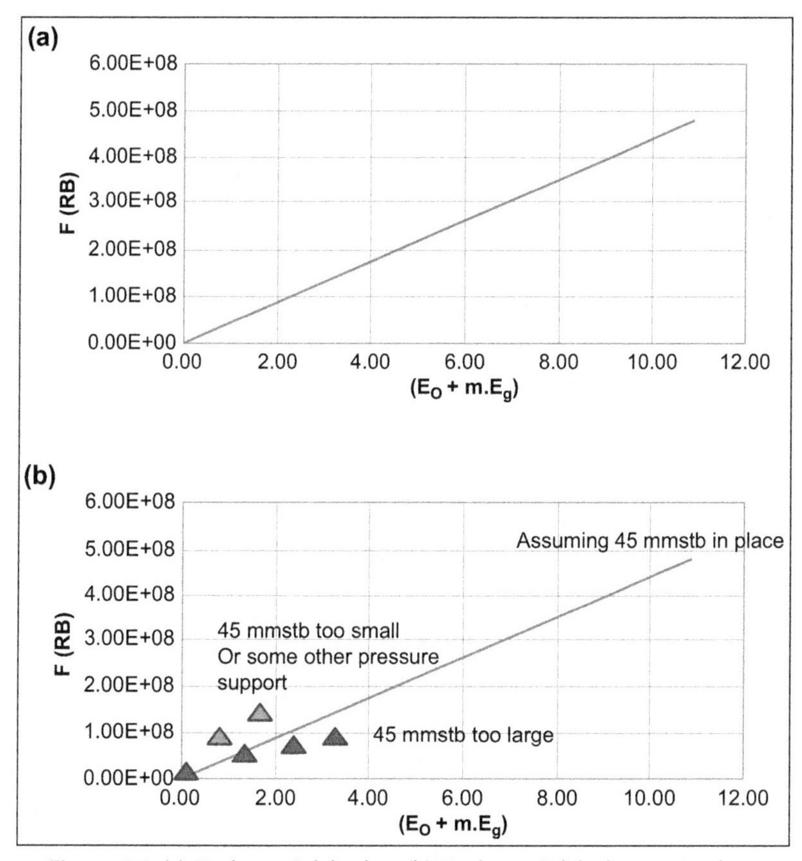

Figure 4.2 (a) Havlena–Odeh plots (b) Havlena–Odeh diagnostic plots.

If actual data points begin to diverge from the assumed in place curve, the original estimate may be modified (see Fig. 4.2).

4.3 EXTENDING MATERIAL BALANCE EQUATIONS TO OBTAIN PRODUCTION PROFILES

In both gas and oil cases above, we relate cumulative production to pressure decline. It is possible to extend this to relate production to time. This can be done using a simple cylindrical tank model. A spreadsheet on "gas decline" is available to model gas rate decline with time. Results from this are approximate, but very useful in first estimates of potential single-well performance.

4.3.1 Dry Gas Rate Decline With Time

The dry gas rate with time is calculated as follows, starting from Darcy's law in cylindrical coordinates (see Fig. 2.7a):

$$q = kh\frac{p^2 - p_w^2}{1422\mu ZT\left[\ln\left(\frac{r_e}{r_w}\right) - 1/2\right]}(q \text{ in scf/d}) \qquad [4.10]$$

$$[q(cf/d) = 14.7\cdot q(\text{scf}/d)\cdot T(°R)/(p\cdot T_o(°R))q(\text{moles}/d) = p\cdot q(cf/d)/ZRT]$$

$$n(t + \Delta t) = n(t) - q(t)\cdot t \qquad [4.11]$$

$$p = ZRT/V_{tank} \qquad [4.12]$$

From the starting conditions, assuming p_w = bottom hole (drawdown) pressure (p_o), p = initial reservoir pressure, and $n(t = 0)$ the initial number of moles in the tank, $q(t)$ is calculated from Eq. [4.10], from which $n(t + \Delta t)$ can be calculated (Eq. [4.11]) and from this $p(t + \Delta t)$ (the average tank pressure at time $t + \Delta t$) can be determined using Eq. [4.12].

This is then continued with time steps Δt.

An example of the use of the spreadsheet on "gas decline" is given in Fig. 4.3, where we show gas rate as a function of time and as a function of cumulative production.

The simplifying assumption here is that we have an average pressure across the whole tank at any time. An assumption of linear change in $(1/\mu Z)$ with pressure is also implicit in Eq. [4.10]. We can have a series of interconnected tanks to give a better representation of the real system but this is obviously a more complicated model and is fully explicit so can be unstable but it can still be useful for some modeling in particular of shale gas wells.

4.3.2 Wet Gas Rates

These are calculated as for dry gas, but with a liquid/gas ratio (l_{gr}) used to calculate the liquid rate:

$$q_{liquid} = q_{gas}\cdot l_{gr} \qquad [4.13]$$

A wet gas example using the spreadsheet "gas decline-zz" is shown in Fig. 4.4.

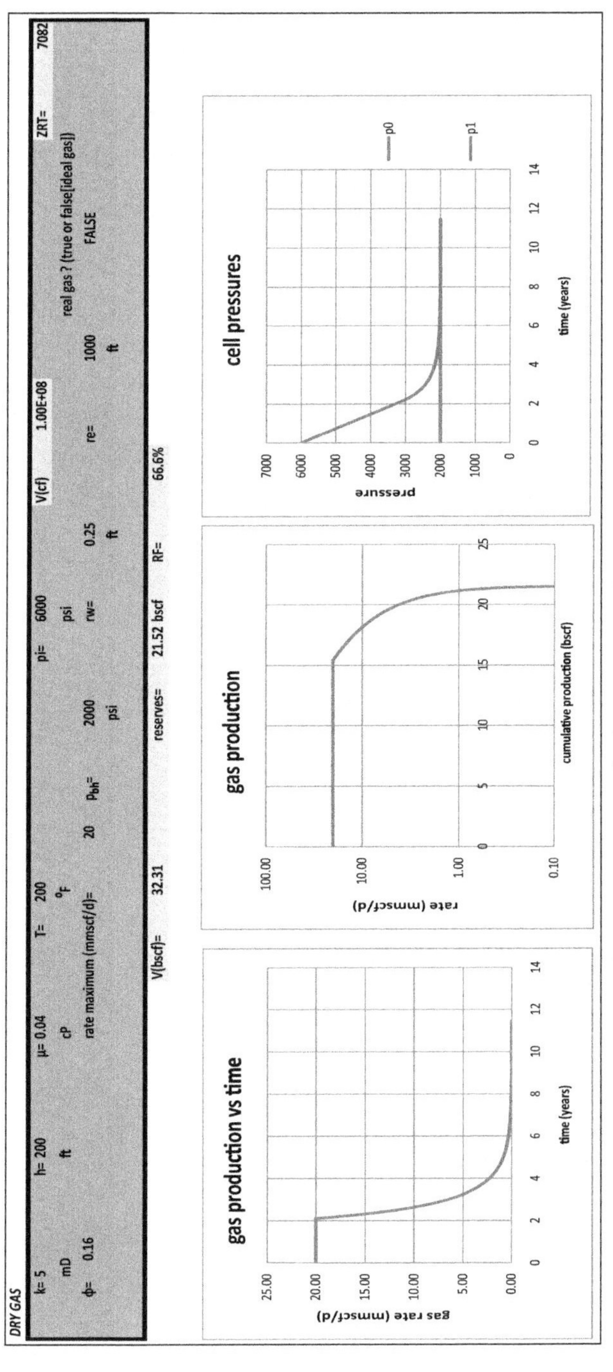

Figure 4.3 Dry gas depletion—example input and output.

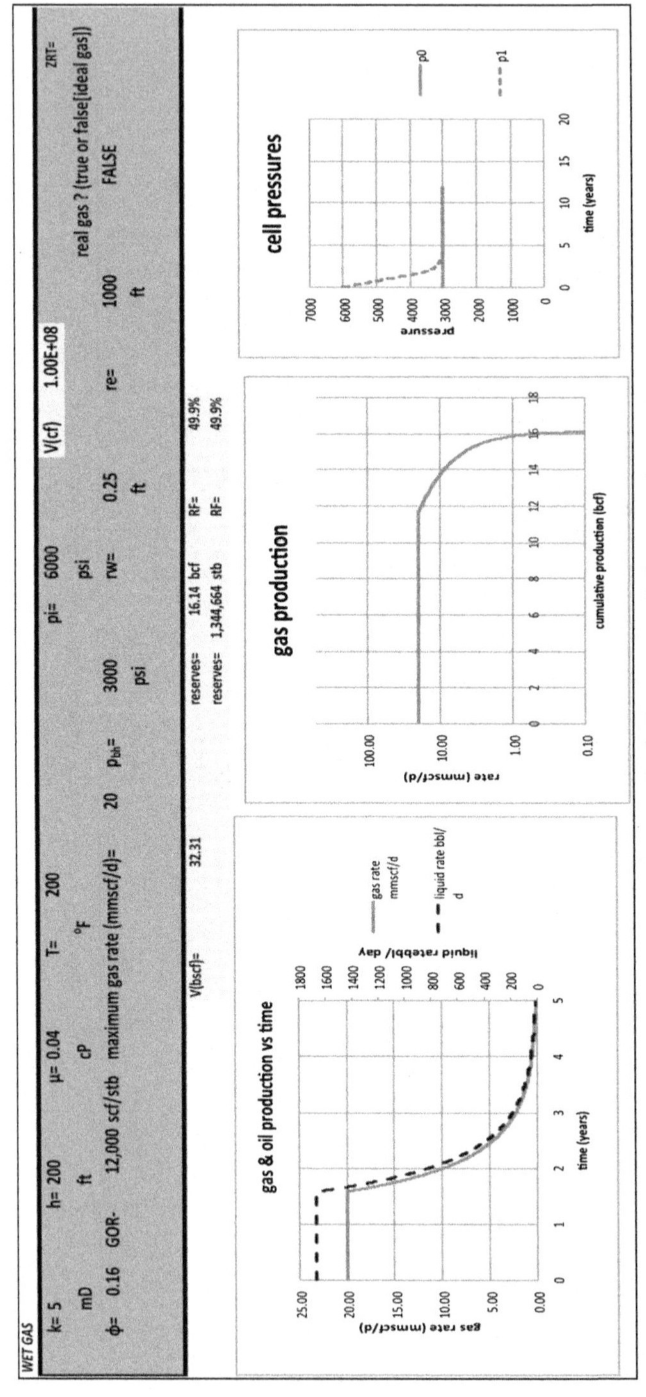

Figure 4.4 Wet gas depletion—example input and output.

4.3.3 Gas Condensate Rates

A separate spreadsheet on "gas condensate" is available for this more complex case.

4.3.3.1 Depletion

For straight depletion this is a simple extension of the above model with the input of a starting produced condensate/gas ratio (or its reciprocal produced GOR). As pressure drops below the dew point, liquid drops out in the reservoir and produced GOR increases so that a gradient of GOR with pressure needs to be input (Fig. 4.5). Field data suggest that the gradient of GOR with pressure is approximately linear. The richer the original reservoir fluid, the steeper the increasing GOR gradient.

A typical result showing gas and liquid rates as pressure decreases is given in Fig. 4.6.

The change in gradient of the liquid rate at dew-point pressure can be seen.

4.3.3.2 Recycling

In this case dry gas from surface separation is reinjected both to maintain reservoir pressure above dew point (as far as possible) and to sweep the richer gas towards the producing well. In terms of our simple radial model, this effective sweep is shown in Fig. 4.7. This model assumes a self-sharpening front, giving effectively piston-type displacement of rich gas by injected gas. However, a sweep efficiency factor can be input. The number of moles present at time $t + \Delta t$ is modified to allow for reinjected gas:

$$n(t + \Delta t) = n(t) - q(t)[1 - f(t)]t \qquad [4.14]$$

where $f =$ the fraction of produced gas volume q replaced by injected gas. Where recycled gas is injected, reservoir pressure is maintained, depending on the fraction of produced gas reinjected.

Gas sales are calculated after removal of reinjected gas and liquids produced, and gas sold is calculated using the condensate/gas ratio, as with the depletion cases. A blowdown period follows complete sweep of the richer gas. An example (with a 50% replace ratio) of the input and output from the spreadsheet "gas condensate" is shown in Fig. 4.8.

Examples with various replacement factors from 0% to 100% are shown in Fig. 4.9.

Liquid recovery increases as we increase the replacement ratio.

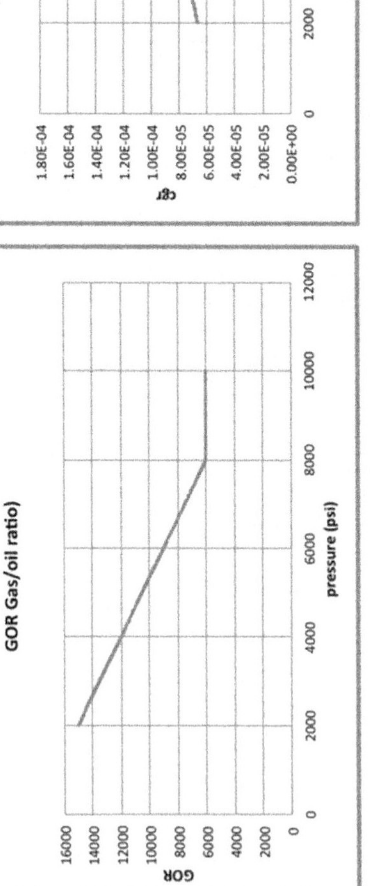

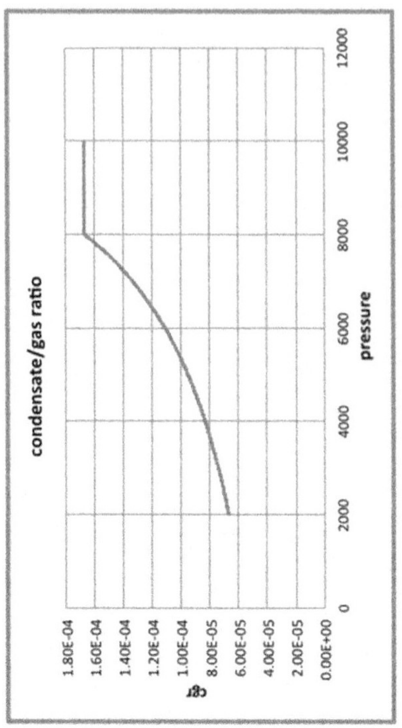

Figure 4.5 Typical changes in produced GOR and condensate/gas ratio with pressure for a gas condensate.

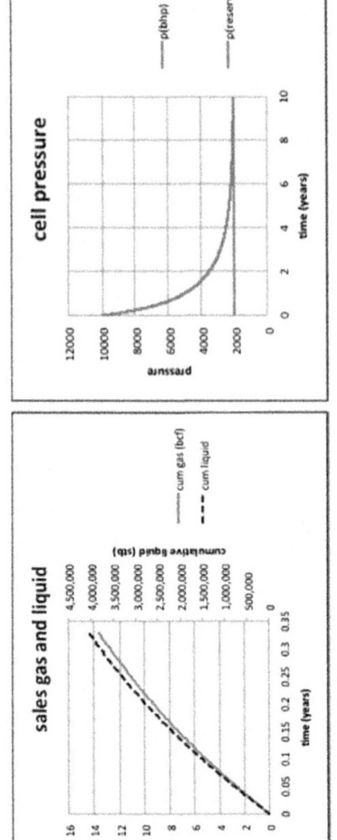

Figure 4.6 Example of gas condensate rates—gas and liquid for straight depletion.

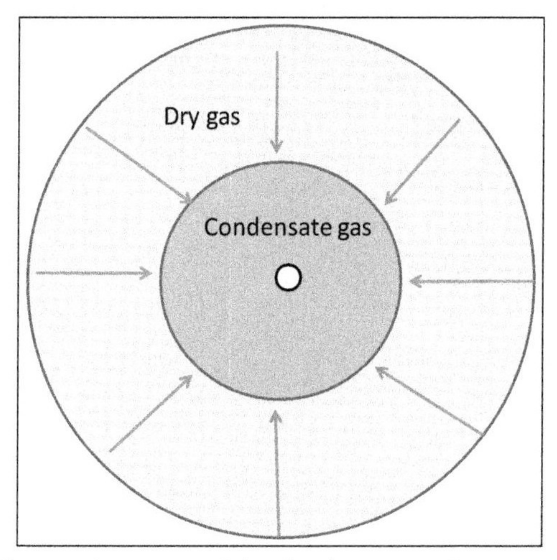

Figure 4.7 Gas condensate model—simple gas condensate sweep model.

4.3.4 Oil Rates With Time

Here we can use a model based on the black oil representation. Determining production q as a function of time requires the Darcy equation:

$$q = \frac{kh(p - p_{\mathrm{w}})}{141.2\mu \ln\left(\frac{r_{\mathrm{e}}}{r_{\mathrm{w}}}\right)}$$

If we neglect gas or water ingress and rock–water expansion effects, from Eq. [4.9] material balance with these conditions then gives

$$\Delta N = N[(B_{\mathrm{o}} - B_{\mathrm{oi}}) + (R_{\mathrm{si}} - R_{\mathrm{s}}) \cdot B_{\mathrm{g}}] \qquad [4.15]$$

It is assumed to be undersaturated throughout the time period covered. We can assume simple linear relationships (Fig. 4.10), so that

$$\text{where } p > p_{\mathrm{b}} \quad \begin{aligned} B_{\mathrm{o}} &= m_1 \cdot (p_{\mathrm{b}} - p) + B_{\mathrm{o}}(p_{\mathrm{b}}) \\ R_{\mathrm{s}} &= R_{\mathrm{si}} \end{aligned} \qquad [4.16]$$

$$\text{where } p < p_{\mathrm{b}} \quad \begin{aligned} B_{\mathrm{o}} &= m_2 \cdot (p - p_{\mathrm{b}}) + B_{\mathrm{o}}(p_{\mathrm{b}}) \\ R_{\mathrm{s}} &= m_3 \cdot (p - p_{\mathrm{b}}) + R_{\mathrm{si}} \end{aligned} \qquad [4.17]$$

and for gas,

$$\mathrm{B_g = n_1/p = 5.044/(PZT)} \qquad [4.18]$$

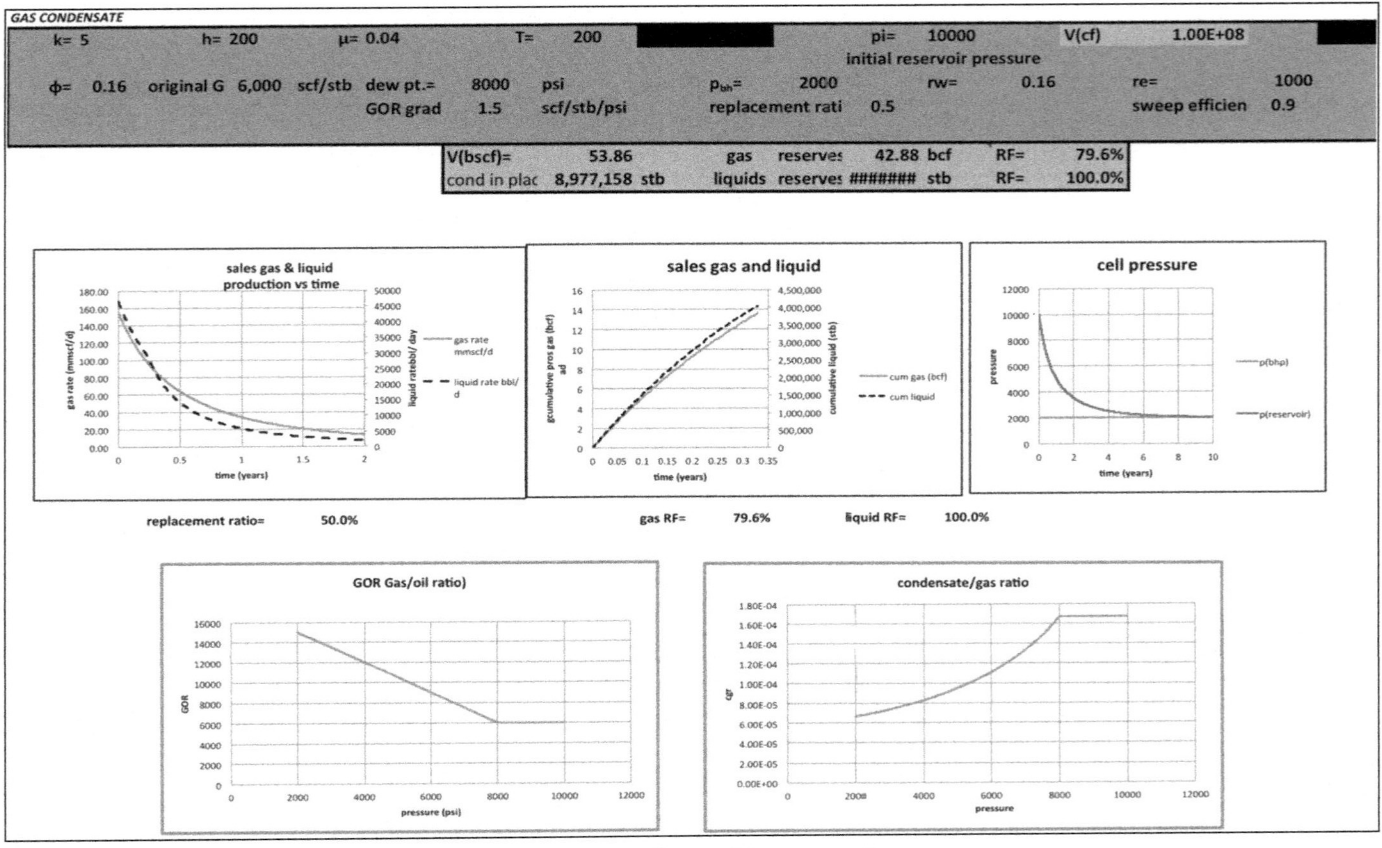

Figure 4.8 Excel spreadsheet example.

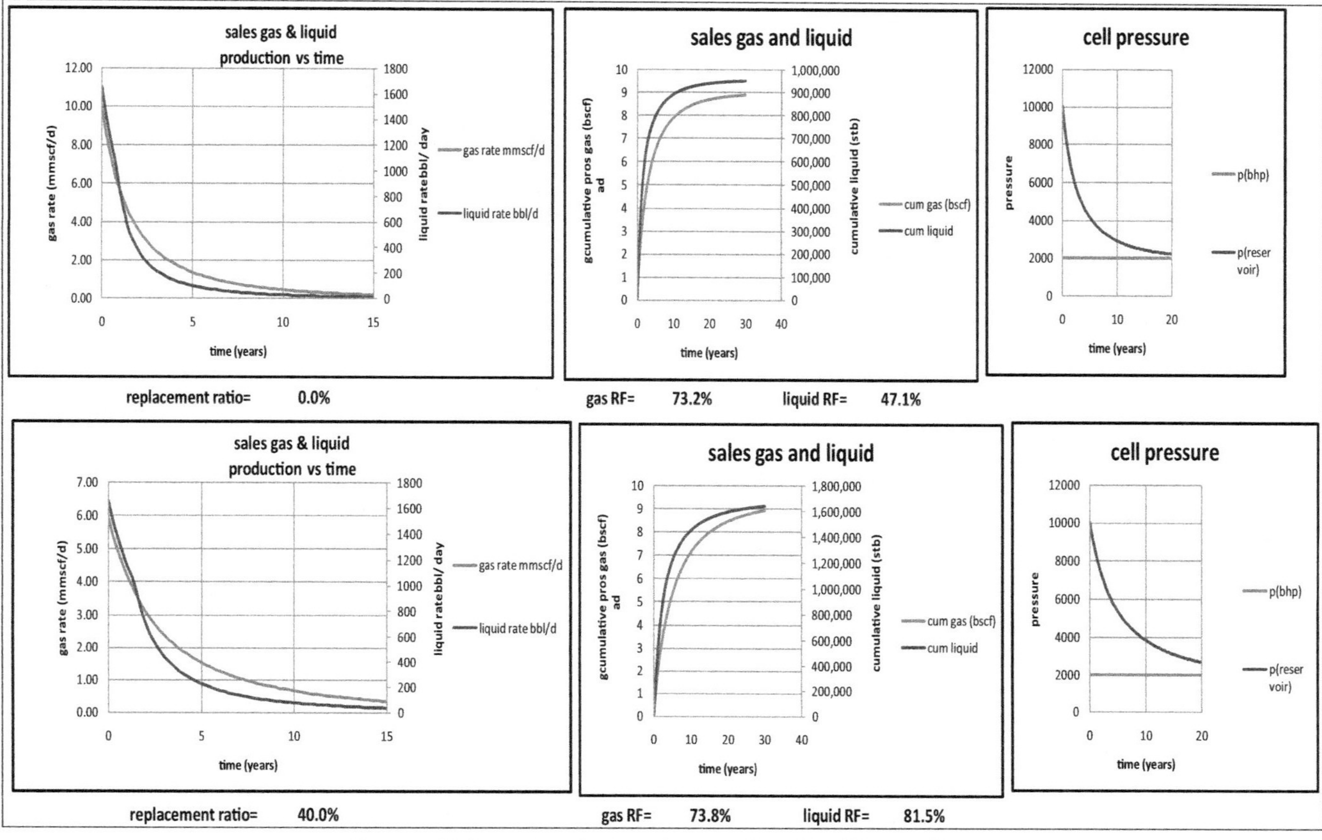

Figure 4.9 Example of rates of gas and liquid with recycling.

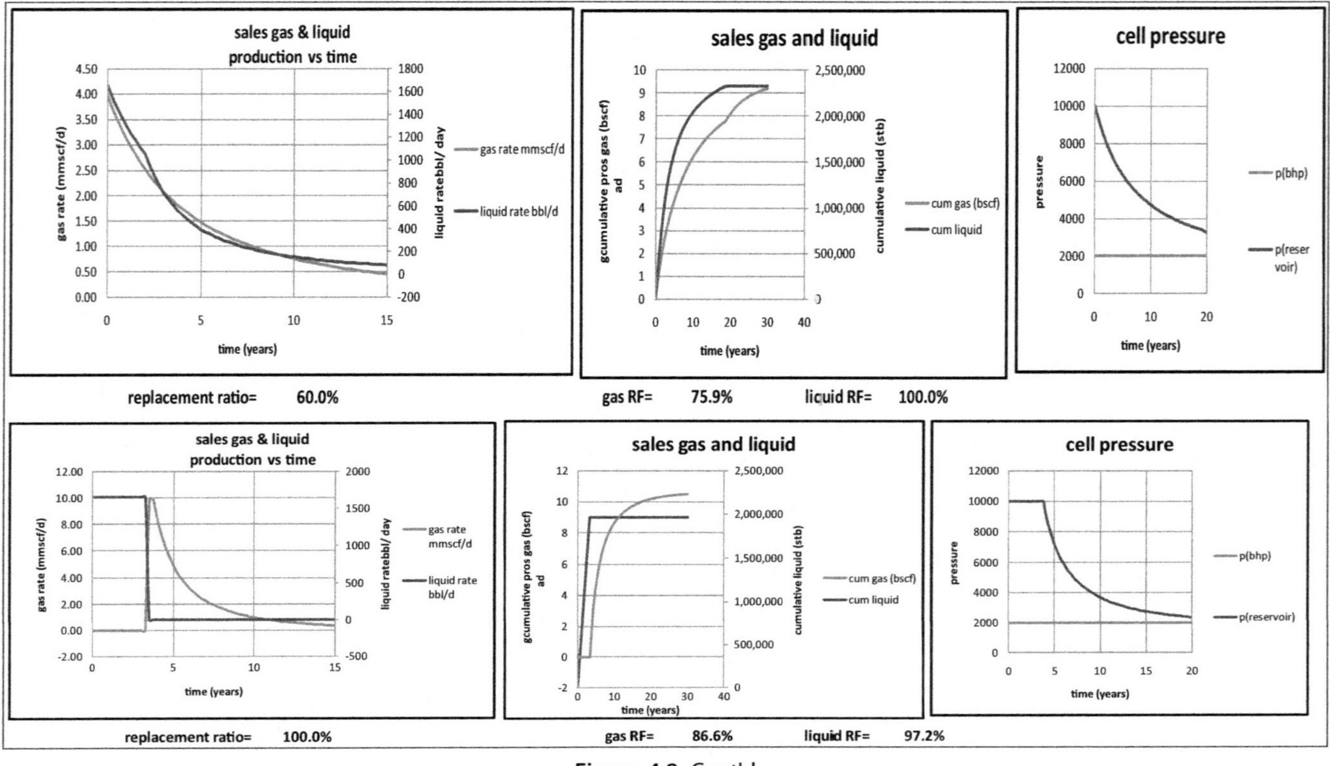

Figure 4.9 Cont'd

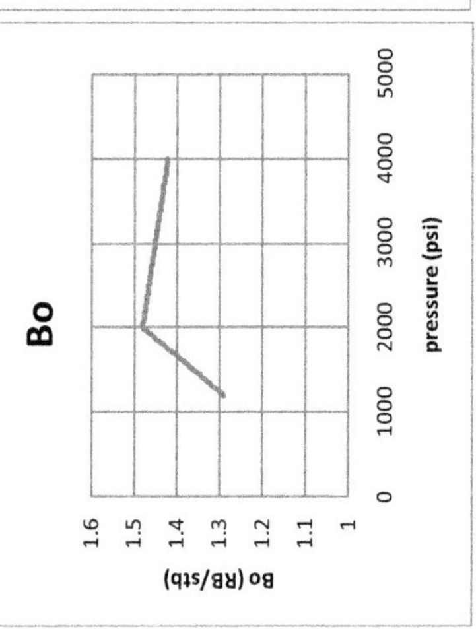

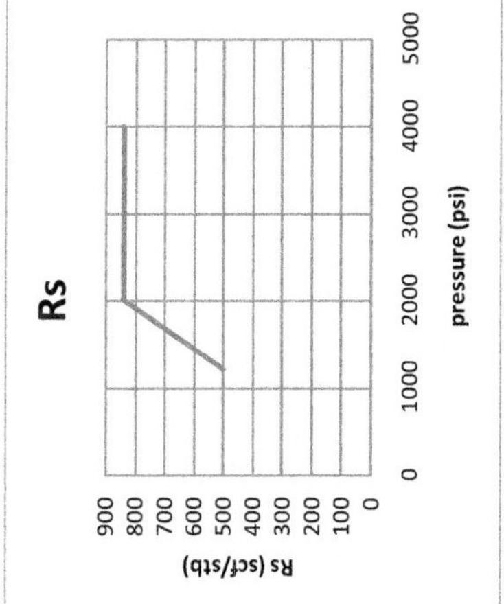

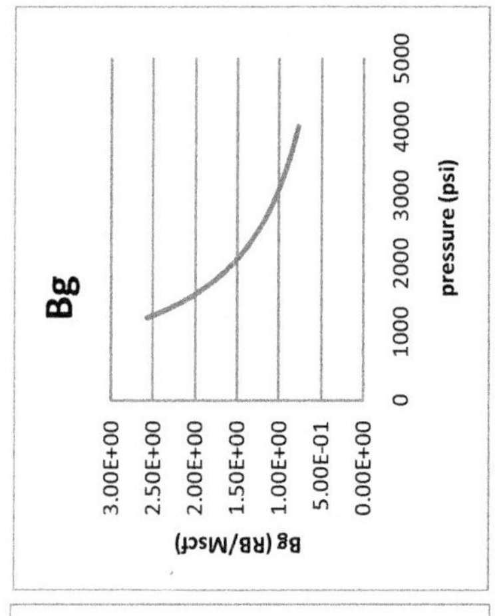

Figure 4.10 Example black oil functions in model.

An Excel spreadsheet (solution gas drive) based on the above is available. Details of the solution algorithm are given in appendix "Simple Oil Material Balance for Rate as a Function of Time."

An example of output from this model is shown in Fig. 4.11, where the initial reservoir pressure is 5000 psi and the bubble-point pressure is 4000 psi.

The very limited oil production above bubble point due to the low compressibility of oil is clearly seen.

4.4 WATER–FLOOD PERFORMANCE ESTIMATION FROM ANALYTICAL EQUATIONS

4.4.1 Frontal Advance Equations

If we assume no mass transfer between phases, incompressible fluids and a homogeneous system (as shown in Fig. 4.12), then if q_w = flow rate of water; q_o = flow rate of oil; ρ_w = density of water; ρ_o = density of oil; S_w = water saturation; and φ = porosity:

$$(q_w\rho_w)_x - (q_w\rho_w)_{x+\Delta x} = A\varphi\frac{\partial}{\partial t}(S_w\rho_w)\cdot\Delta x \qquad [4.19]$$

Mass of water entering − mass leaving = mass accumulation rate of water,

$$(q_w\rho_w)_x - (q_w\rho_w)_{x+\Delta x} = -\Delta(q_w\rho_w)/\Delta x = \frac{\partial}{\partial x}(q_w\rho_w) \qquad [4.20]$$

(as Δx goes to 0)

therefore,

$$\frac{\partial}{\partial x}(q_w\rho_w) = -A\varphi\frac{\partial}{\partial t}(S_w\rho_w) \qquad [4.21]$$

But the density of water (ρ_w) is assumed constant, so that

$$\frac{\partial}{\partial x}(q_w) = -A\varphi\frac{\partial}{\partial t}(S_w) \qquad [4.22]$$

Rearranging, we get

$$\left(\frac{\partial S_w}{\partial t}\right)x = -\frac{1}{A\varphi}\left(\frac{\partial q_w}{\partial x}\right)t \qquad [4.23]$$

We can consider q_w as a function of saturation S_w, so that

$$\left(\frac{\partial S_w}{\partial t}\right)x = -\frac{1}{A\varphi}\left(\frac{\partial q_w}{\partial S_w}\right)t\cdot\left(\frac{\partial S_w}{\partial x}\right)t \qquad [4.24]$$

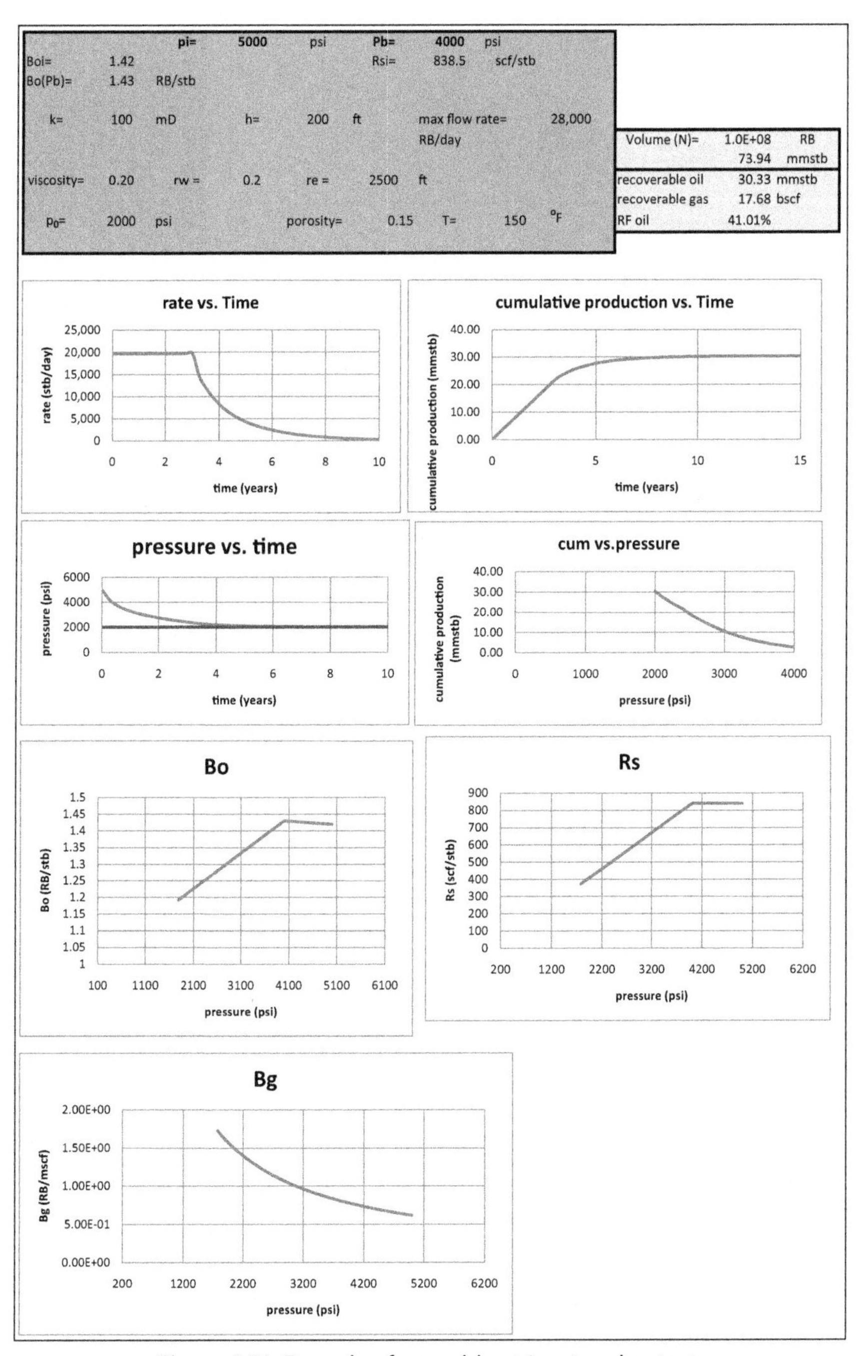

Figure 4.11 Example of spreadsheet input and output.

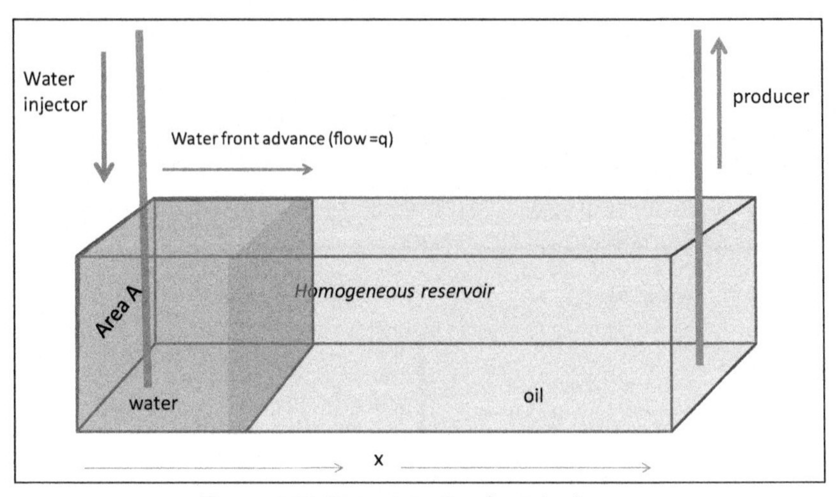

Figure 4.12 Water injection frontal advance.

$$S_w = f(x, t)$$

Therefore,

$$dS_w = \left(\frac{\partial S_w}{\partial x}\right)t + \left(\frac{\partial S_w}{\partial t}\right)x \quad [4.25]$$

for a constant $S_w \cdot dS_w = 0$

$$\left(\frac{\partial S_w}{\partial x}\right)t = -\left(\frac{\partial S_w}{\partial t}\right)x \quad [4.26]$$

differentiating with respect to t with S_w constant,

$$\left(\frac{\partial S_w}{\partial x}\right)t \cdot \left(\frac{\partial x}{\partial t}\right)S_w = -\left(\frac{\partial S_w}{\partial t}\right)x \quad [4.27]$$

rearranging, we get,

$$\left(\frac{\partial x}{\partial t}\right)S_w = \frac{1}{A\varphi}\left(\frac{\partial q_w}{\partial S_w}\right)t \quad [4.28]$$

We now define:

$$f_w = \frac{q_w}{q_w + q_t} \quad [4.29]$$

The fractional flow of water, where q_t = total flow = $(q_w + q_o)$,

$$q_w = q_t f_w \quad [4.30]$$

Since fluids are incompressible q_t a constant with respect to S_w:

$$\left(\frac{\partial x}{\partial t}\right) S_w = \frac{q_t}{A\varphi}\left(\frac{\partial f_w}{\partial S_w}\right) t \quad [4.31]$$

This is the Buckley—Leverett equation, a key equation giving the rate of advance of a given water saturation front as a function of total flow rate and the derivative of f_w with water saturations. The key term here is the fractional flow rate of water f_w, since,

$$u_x = -\frac{k}{\mu}\cdot\frac{d\varphi}{dx}$$

and

$$\phi = p + \rho g D \quad [4.32]$$

where ϕ = flow potential; ρ = density; and D = depth.

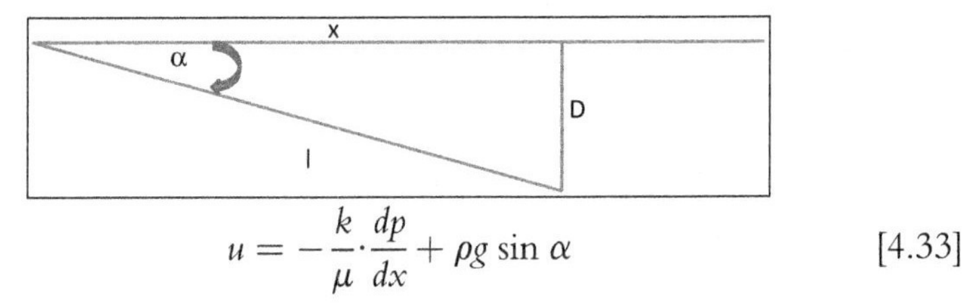

$$u = -\frac{k}{\mu}\cdot\frac{dp}{dx} + \rho g \sin\alpha \quad [4.33]$$

Some manipulation then gives

$$f_w = u_w/(u_w + u_o) \quad [4.34]$$

and

$$f_w = \frac{\left(1 + \left(u\frac{kk_{rw}}{u_t\mu_o}\right) g\Delta\rho \sin\alpha\right)}{1 + \frac{\mu_w k_{ro}}{\mu_o k_{rw}}} \quad [4.35]$$

or in field units

$$f_w = \frac{1 - 0.001127kk_{ro}A[0.4335(\gamma_w - \gamma_o)\sin(\alpha)]/q_T\mu_o}{1 + \frac{\mu_w k_{ro}}{\mu_o k_{rw}}} \quad [4.36]$$

where A = cross-sectional area (reservoir thickness × reservoir width); α = dip angle; Υ_w = specific gravity of water; and Υ_o = specific gravity of oil. For horizontal flow:

$$f_w = \frac{1}{1 + \frac{\mu_w k_{ro}}{\mu_o k_{rw}}} \quad [4.37]$$

The Buckley–Leverett equation can be integrated with respect to time to give

$$x(S_w) = \left(\frac{df_w}{dS_w}\right) S_w \qquad [4.38]$$

$Q(t)$ = total fluid injected at time t.

$\frac{df_w}{dS_w}$ is a function of relative permeability and viscosity ratios. Typical water and oil relative permeabilities are shown in Fig. 4.13a.

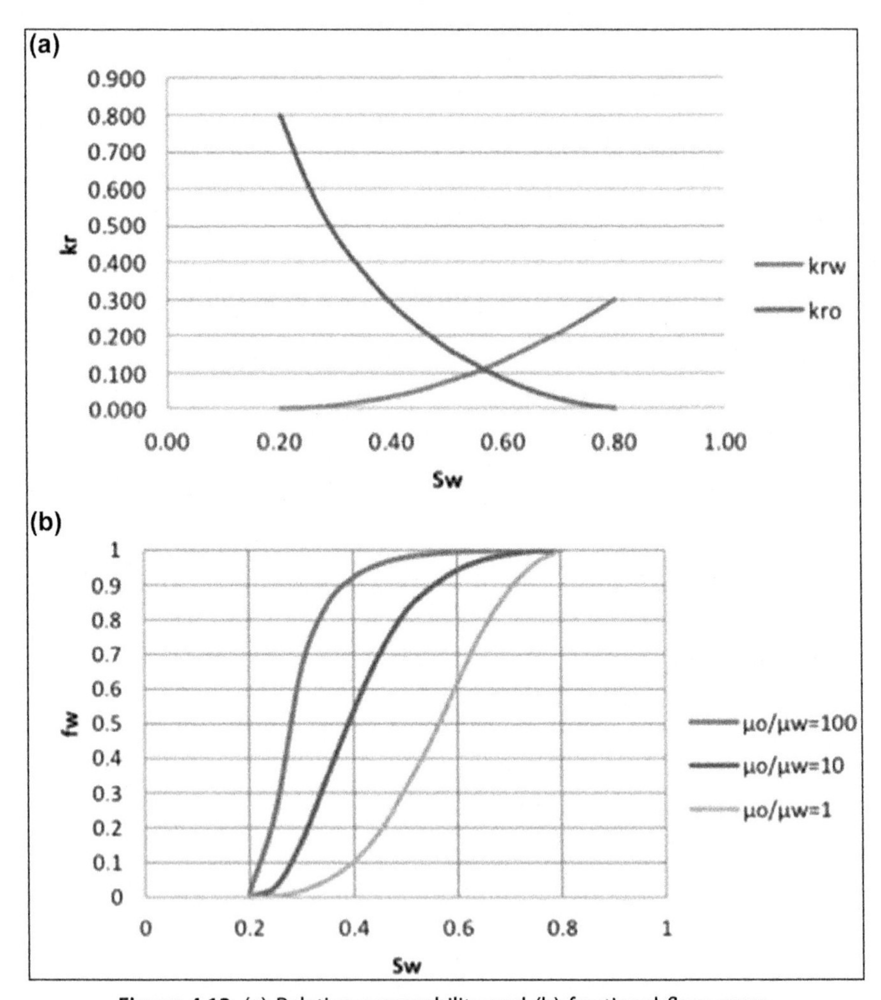

Figure 4.13 (a) Relative permeability and (b) fractional flow curve.

Using the above relative permeabilities for a range of oil/water viscosity ratios, we obtain the fractional flow curves shown in Fig. 4.13b. Higher oil/water viscosity ratios give steeper f_w curves, which result in poorer oil recovery.

Derivatives of water fractional flow with respect to water saturation are shown in Fig. 4.14 for the same range of viscosity ratios.

The velocity of a given water saturation is proportional to the derivative $\frac{df_w}{dS_w}$.

The full curve is physically unrealistic, as we cannot have two water saturations at one point. What actually happens is that a shock front develops, as discussed below.

4.4.1.1 Piston-like Displacement

If we consider the position at the water injection well, for piston-like displacement there is no mixing of oil and water, and this sharp front advances as water continues to be injected (Fig. 4.15). This would be an ideal situation, with all mobile oil displaced towards the production well.

4.4.1.2 Self-Sharpening Systems

If we consider Fig. 4.18 (the Buckley—Leverett equation), it can be seen that the full curves are physically unrealistic since we cannot have two water saturations at one point. What actually happens is that the higher

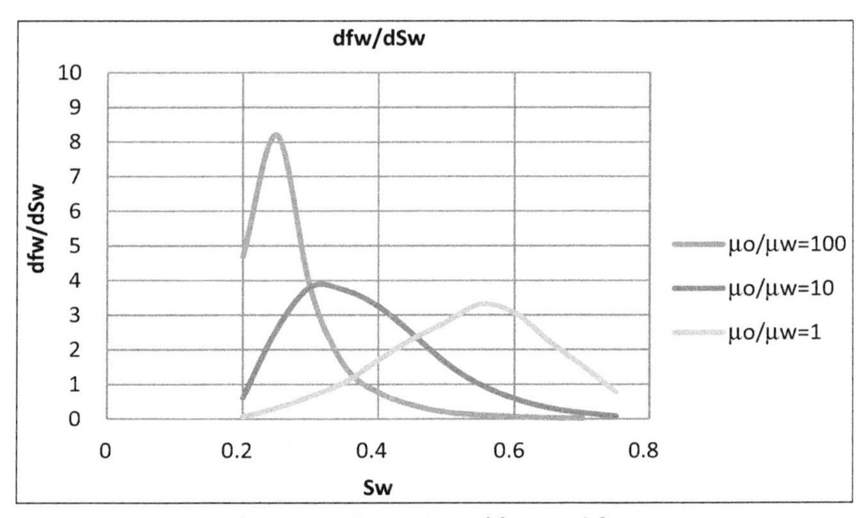

Figure 4.14 Derivatives of fractional flow.

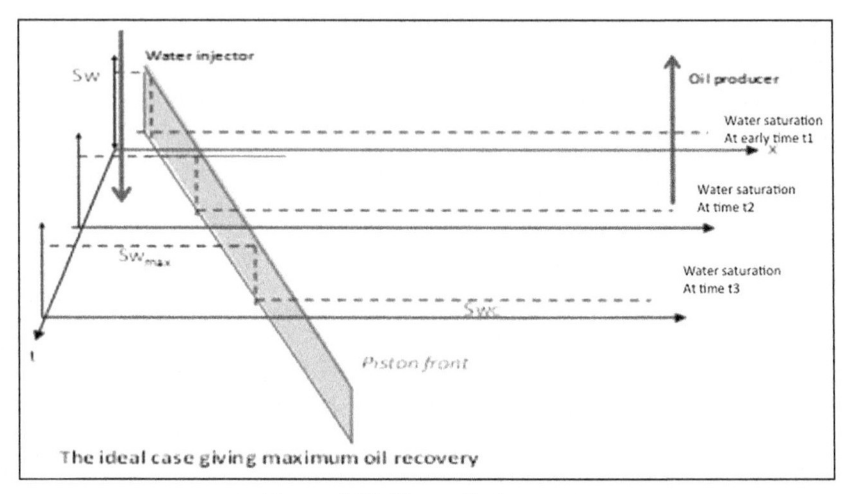

Figure 4.15 Piston displacement.

water front rates at intermediate water saturations overtake the lower water saturation fronts to give a self-sharpening front (Fig. 4.16), resulting in a shock front developing. It can be seen in Fig. 4.16 that this occurs at higher water saturation for systems with low oil/water viscosity, which therefore have better recovery of oil than where we have high oil/water viscosity ratios.

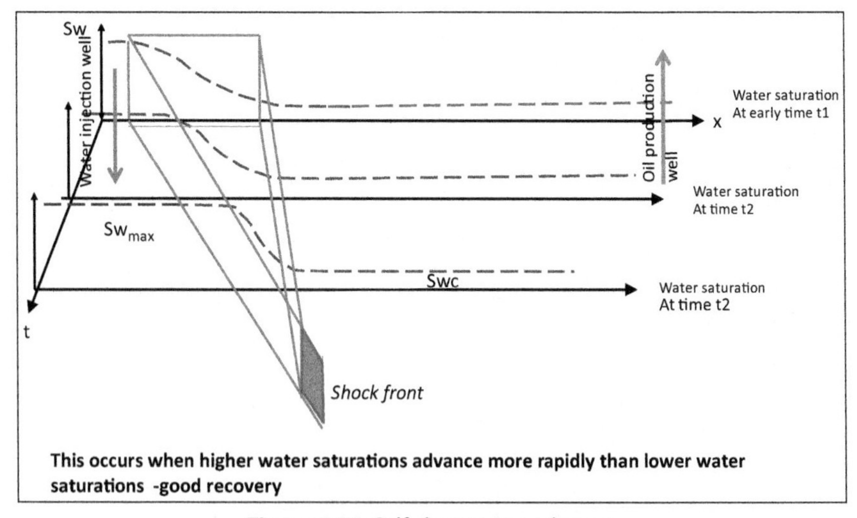

Figure 4.16 Self-sharpening advance.

4.4.1.3 Nonsharpening Systems

High values of $\frac{df_w}{dS_w}$ occur at lower water saturations (in particular for high oil viscosity systems—see Fig. 4.17) and we get a nonsharpening behavior.

In general, therefore, where higher water saturations (following water injection) have higher water advance rates (Fig. 4.18) we have a self-sharpening system, and where higher water saturations (following water injection) have lower water advance rates we have a nonsharpening system (Fig. 4.18).

The applicable part of the curve is therefore that between $S_w = S_{wf}$ and $1 - S_{or}$, where S_{wf} is the shock front advance water saturation.

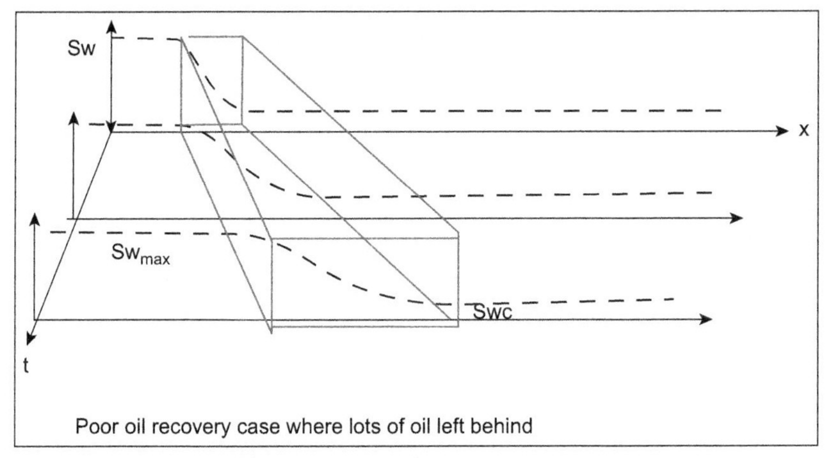

Figure 4.17 Nonsharpening system.

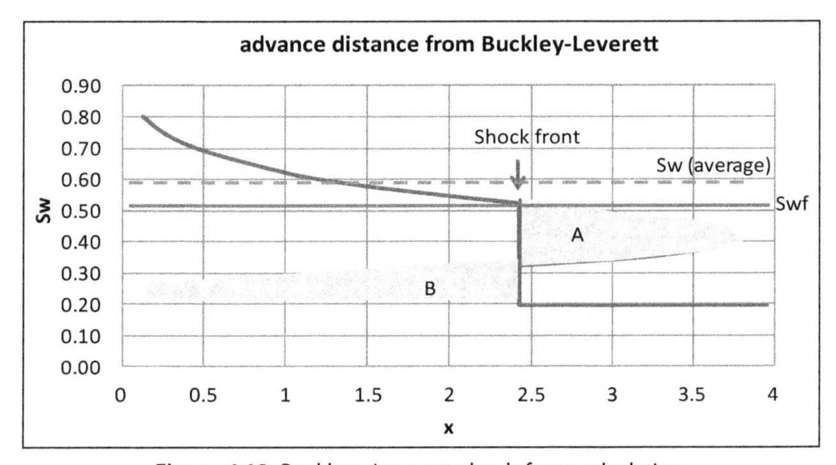

Figure 4.18 Buckley–Leverett shock front calculation.

To ensure mass balance we need to remove equal areas A and B in the Buckley–Leverett plot to determine S_{wf}—see Fig. 4.18.

A better way is that proposed by Welge. Integrating the saturation distribution from $x = 0$ to the shock front, it can be shown that a tangent to the water fractional flow curve will give both S_{wbt} (the water saturation at the shock front) and the average S_w behind the shock front (Fig. 4.19).

$S_{w(average)}$ can be used to determine oil recovery at a given time.

4.4.1.3.1 Steps

1. Draw the fractional flow curve as shown above.
2. Draw tangent as shown.
3. The point of tangency gives S_{wbt}.
4. Extrapolation to $f_w = 1$ gives average water saturation $S_{w(average)} = \overline{S_w}$ behind the shock front at breakthrough time t_{bt}.

4.4.1.3.2 Position of Any Given Water Saturation Front S_w

$$x_{S_w} = \frac{Q(t)}{A\phi} \cdot \left(\frac{df_w}{dS_w}\right)_{S_w} \qquad [4.39]$$

or if flow rate q is a constant with time $Q(t) = q \cdot t$:

$$x_{S_w} = \left(q\frac{t}{A\varphi}\right) \cdot \left(\frac{df_w}{dS_w}\right)_{S_w} \qquad [4.40]$$

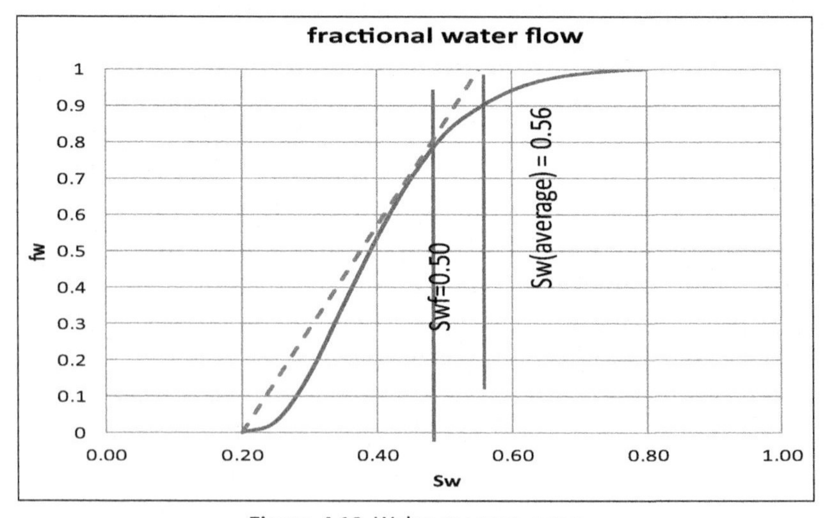

Figure 4.19 Welge tangent curve.

or in field units:

$$x_{S_w} = \left(5.615q\frac{t}{A\varphi}\right)\cdot\left(\frac{df_w}{dS_w}\right)_{S_w} \quad [4.41]$$

where $q =$ a constant water injection rate, and q is in bbl/day, t is in days and A is in ft^2.

4.4.2 Time to Water Breakthrough

Since up to water breakthrough time (t_{bt}) no water is produced, total water volume injected = water volume change within porous volume, so that

$$q_{inj}t_{bt} = \varphi Al\left(\overline{S_w} - S_{wc}\right) \quad [4.42]$$

or

$$t_{bt} = \frac{\varphi Al\left(\overline{S_w} - S_{wc}\right)}{q} \quad [4.43]$$

where $\overline{S_w} =$ the average water saturation behind the breakthrough front (from tangent plot) in field units

$$t_{bt} = \frac{\varphi Al\left(\overline{S_w} - S_{wc}\right)}{5.615q} \quad [4.44]$$

4.4.3 Sweep Efficiency and Recovery Factor at Breakthrough

Sweep efficiency at breakthrough is given by

$$E = \frac{\overline{S_w} - S_{wc}}{1 - S_{wc}} \quad [4.45]$$

Recovery factor at breakthrough

$$RF = \overline{S_w} - S_{wc} \quad [4.46]$$

A measure of the likely sweep efficiency is given by the "mobility ratio": the ratio of mobility of the displacing fluid to the fluid being displaced. It is dimensionless and given by:

$$M = \frac{k_w(S_{or})/\mu_w}{k_o(S_{wc})/\mu_o} \quad [4.47]$$

The value of M gives a measure of the expected sweep efficiency. If the displacing phase is more mobile than the displaced phase, this is an unfavorable situation for oil recovery. We can assume that if

$M > 1$ = unfavorable displacement;

$M < 1$ = favorable displacement.

4.4.4 Production Rates

Up to water breakthrough only oil is produced, so if reservoir pressure is kept constant by water injection, the oil production rate is equal to the water injection rate (assuming for simplicity that oil and water densities are approximately equal under reservoir conditions). At the point of breakthrough water saturation will rise from 0 to S_{wbt}, so produced oil saturation will decrease from 1.0 to $(1 - S_{wbt})$ and oil production rate will decrease correspondingly.

Following water breakthrough, the oil rate will decline as the water saturation increases. This can be estimated numerically with time using the top part of the fractional flow curve shown in Fig. 4.20.

This relates fractional flow of water to water saturation, and we need to relate it to time. This can be done by using incremental steps in S_w and the

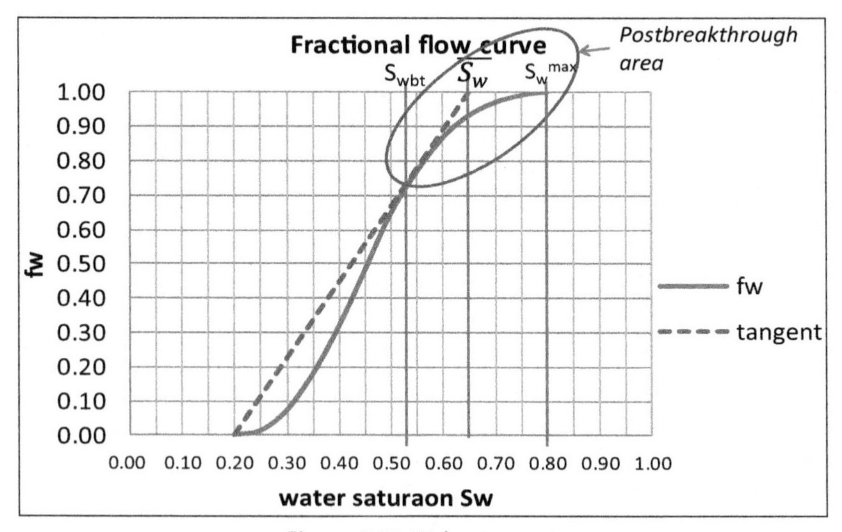

Figure 4.20 Welge tangent.

derivative of S_w with time. Available software (waterflood) has been used to generate the example shown in Fig. 4.21.

If we consider the approximate nature of the Buckley—Leverett/Welge approach, it may be more appropriate to use a simple decline for this postbreakthrough period. Where we have a high water-cut at breakthrough, exponential decline ($b = 0$) is realistic; however, when we have a lower breakthrough in water saturation, slower harmonic decline ($b = 1$) is reasonable to use. This can be done in the available spreadsheets (waterflood-and aggregation-oil—see chapter: Field Appraisal and Development Planning), where relative permeabilities, viscosities, etc. are input, a tangent can be fitted to the resulting fractional flow curve and S_{wbt}, $\overline{S_w}$, and S_{wc} read off. These can be used with the above equation for time to breakthrough for subsequent input into the "aggregation-vs" spreadsheet to obtain a production profile (Fig. 4.21).

It should be understood that recovery factors calculated using analytical methods like this are "ideal," and the reality will often be significantly less due to heterogeneity in the reservoir leading to "figuring" of water advance.

4.4.5 Excel Spreadsheet "Waterflood"

An example of the input and output is shown in Fig. 4.22.

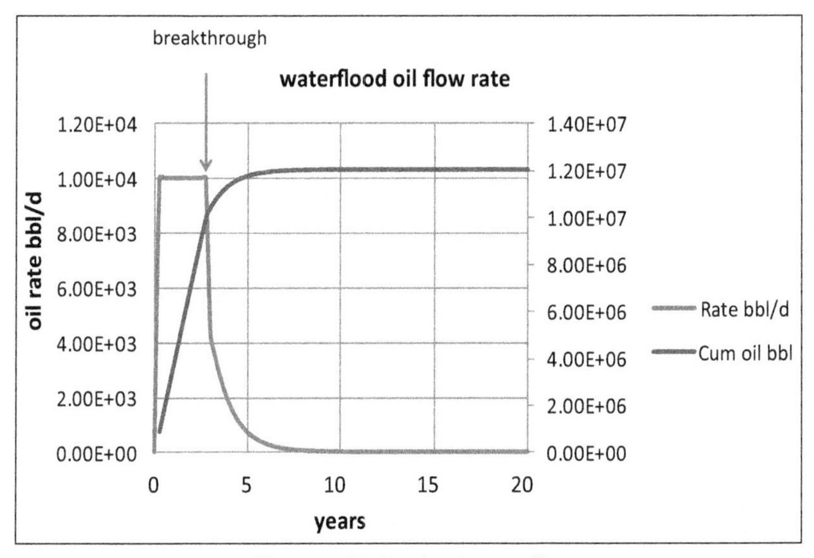

Figure 4.21 Production profile.

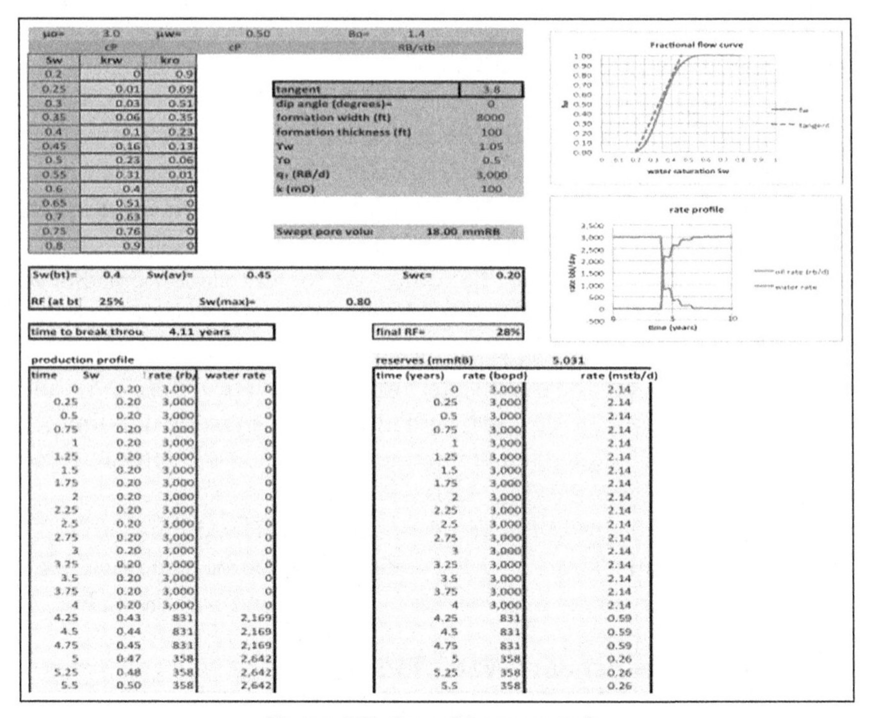

μo=	3.0	μw=	0.50	Bo=	1.4
	cP		cP		RB/stb

Sw	krw	kro
0.2	0	0.9
0.25	0.01	0.69
0.3	0.03	0.51
0.35	0.06	0.35
0.4	0.1	0.23
0.45	0.16	0.13
0.5	0.23	0.06
0.55	0.31	0.01
0.6	0.4	0
0.65	0.51	0
0.7	0.63	0
0.75	0.76	0
0.8	0.9	0

tangent	3.8
dip angle (degrees)=	0
formation width (ft)	8000
formation thickness (ft)	100
Yw	1.05
Yo	0.5
q_t (RB/d)	3,000
k (mD)	100

Swept pore volu 18.00 mmRB

Sw(bt)= 0.4 Sw(av)= 0.45 Swc= 0.20

RF (at bt 25% Sw(max)= 0.80

time to break throu 4.11 years

final RF= 28%

production profile

time	Sw	rate (rb/	water rate
0	0.20	3,000	0
0.25	0.20	3,000	0
0.5	0.20	3,000	0
0.75	0.20	3,000	0
1	0.20	3,000	0
1.25	0.20	3,000	0
1.5	0.20	3,000	0
1.75	0.20	3,000	0
2	0.20	3,000	0
2.25	0.20	3,000	0
2.5	0.20	3,000	0
2.75	0.20	3,000	0
3	0.20	3,000	0
3.25	0.20	3,000	0
3.5	0.20	3,000	0
3.75	0.20	3,000	0
4	0.20	3,000	0
4.25	0.43	831	2,169
4.5	0.44	831	2,169
4.75	0.45	831	2,169
5	0.47	358	2,642
5.25	0.48	358	2,642
5.5	0.50	358	2,642

reserves (mmRB) 5.031

time (years)	rate (bopd)	rate (mstb/d)
0	3,000	2.14
0.25	3,000	2.14
0.5	3,000	2.14
0.75	3,000	2.14
1	3,000	2.14
1.25	3,000	2.14
1.5	3,000	2.14
1.75	3,000	2.14
2	3,000	2.14
2.25	3,000	2.14
2.5	3,000	2.14
2.75	3,000	2.14
3	3,000	2.14
3.25	3,000	2.14
3.5	3,000	2.14
3.75	3,000	2.14
4	3,000	2.14
4.25	831	0.59
4.5	831	0.59
4.75	831	0.59
5	358	0.26
5.25	358	0.26
5.5	358	0.26

Figure 4.22 Spreadsheet example.

4.5 QUESTIONS AND EXERCISES

Q4.1. **1.** If an undersaturated reservoir has stock tank oil initially in place of 200 million bbl and we ignore any aquifer influx and pore volume changes, calculate the oil volume in bbl that we might recover with an initial pressure of 2000 psi and a final pressure of 800 psi using the following data: Oil formation volume factor = 1.467 RB/stb at initial reservoir pressure and 1.278 RB/stb at final pressure, Solution gas oil ratio = 834 scf/stb at intial pressure and 464 scf/stb at final pressure. Assume B_g = 0.004 RB/scf and R_p (produced GOR) = 800 scf/stb.

2. What is the theoretical recovery factor and what may reduce this recovery?

Q4.2. If we have reservoir of a 200 bscf (surface conditions) gas field with an initial pressure of 3000 psi and a final pressure of 1000 psi, temperature = 150°F, what volume of gas will be recovered? Use the compressibility plot shown in Fig. 2.40b.

Q4.3. The table below shows oil and water relative permeability. If water viscosity is 0.5 cP and oil viscosity is 2 cP, then for a horizontal system use the software provided ("waterflood") and a Welge tangent to estimate the water saturation at breakthrough and the average water saturation behind the water front at this point. Calculate the oil recovery factor at breakthrough.

Q4.4. Write the Buckley–Leverett equation for the advance of a given water saturation with time.

S_w	K_{rw}	k_{ro}
0.20	0.000	0.800
0.25	0.002	0.610
0.30	0.009	0.470
0.35	0.020	0.370
0.40	0.033	0.285
0.45	0.051	0.220
0.50	0.075	0.163
0.55	0.100	0.120
0.60	0.132	0.081
0.65	0.170	0.050
0.70	0.208	0.027
0.75	0.251	0.010
0.80	0.300	0.000

Q4.5. Use the relative permeability table from Q4.3. If water viscosity is 0.5 cP and oil viscosity is 5.0 cP, for a horizontal system use the software provided ("waterflood") and a Welge tangent to estimate time to breakthrough, the water saturation at breakthrough and the average water saturation behind the water front at this point. Calculate the oil recovery factor at breakthrough. Assuming the volume of the swept area is 18 mm bbl and the water injection rate is 10,000 RB/day, give the final recovery factor. Show a plot of the oil rate with time.

Q4.6. The tables below show an oil and water relative permeability for oil wet and water wet systems. If water viscosity is 0.5 cP and oil viscosity is 5.0 cP, for a horizontal system use the software provided ("waterflood") and a Welge tangent to estimate the water saturation at breakthrough, the average water saturation behind the water front, time to water breakthrough at this point and final recovery factor.

water wet system

S_w	K_{rw}	K_{ro}
0.20	0.00	0.90
0.25	0.00	0.76
0.30	0.01	0.63
0.35	0.03	0.51
0.40	0.04	0.40
0.45	0.07	0.31
0.50	0.10	0.23
0.55	0.14	0.16
0.60	0.18	0.10
0.65	0.23	0.06
0.70	0.28	0.03
0.75	0.34	0.01
0.80	0.40	0.00

Oil wet system

S_w	K_{rw}	K_{ro}
0.20	0.00	0.90
0.25	0.01	0.69
0.30	0.03	0.51
0.35	0.06	0.35
0.40	0.10	0.23
0.45	0.16	0.13
0.50	0.23	0.06
0.55	0.31	0.01
0.60	0.40	0.00
0.65	0.51	0.00
0.70	0.63	0.00
0.75	0.76	0.00
0.80	0.90	0.00

Q4.7. Use the software "gas decline" to model a dry gas well, assuming permeability = 8 mD, completion length = 140 ft, gas viscosity = 0.04 cP, and porosity = 0.16. We assume an initial reservoir pressure of 6500 psi, a depletion radius of 1000 ft and a wellbore radius of 0.25 ft. Bottom hole pressure is taken as 2000 psi and reservoir temperature as 200°F.

Calculate the production profile, cumulative production, reserves and recovery factor.

Q4.8. Use the software "gas decline" to model a wet gas well, assuming all data as in Q4.7 with a GOR of 12,000 scf/stb.

Calculate the production profiles, cumulative production, reserves and recovery factors.

Q4.9. Use the software "solution gas drive-zz" to model an oil well, assuming the following.

B_{oi} = 1.42 (RB/stb), B_o(Pb) = 1.43 (RB/stb), R_{si} = 838.5 scf/stb.

Initial reservoir pressure = 5000 psi, bubble-point pressure = 3000 psi.

Permeability = 50 mD, completion thickness = 150 ft, oil viscosity = 0.2 cP.

Drainage radius = 2000 ft, wellbore radius = 0.25 ft.

Bottom hole pressure = 1500 psi, porosity*(1 − S_{wc}) = 0.15, reservoir temperature = 150°F.

Calculate the production profile, cumulative production, reserves, and recovery factor.

4.6 FURTHER READING

L.P. Dake, Fundamentals of Reservoir Engineering, Elsevier, 1978.

4.7 SOFTWARE

gas decline
gas condensate
solution gas dive
waterflood

CHAPTER 5

Numerical Simulation Methods for Predicting Reservoir Performance

5.1 INTRODUCTION

Modern reservoir engineering is dominated by the use of numerical reservoir simulators, which, with increasing computer speed, have become an increasingly powerful and important tool in understanding and predicting field performance.

5.2 BASIC STRUCTURE OF NUMERICAL MODELS

The understood geological structure is split into hundreds or thousands of discrete grid cells, which can have varying geometry, with assigned rock properties (volume, porosity, net to gross (NTG), permeability, rock compressibility etc.); see Fig. 5.1 These cells are filled with reservoir fluids: gas, oil, and water with defined fluid properties regarding saturation, relative permeability, capillary pressure, PVT (pressure/volume/temperature)

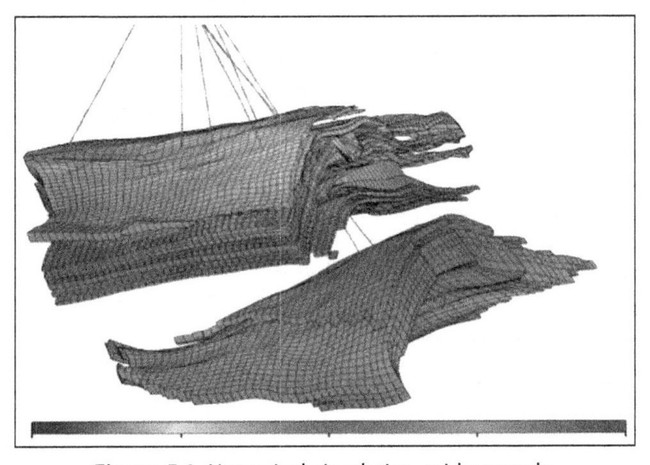

Figure 5.1 Numerical simulation grid example.

Fundamentals of Applied Reservoir Engineering
ISBN 978-0-08-101019-8
http://dx.doi.org/10.1016/B978-0-08-101019-8.00005-3
© 2016 Elsevier Ltd.
All rights reserved.

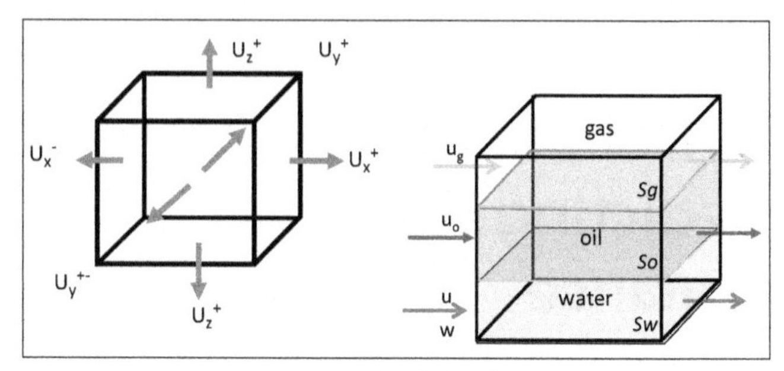

Figure 5.2 Grid cells.

properties, formation volume factors (FVFs), densities, viscosity, etc. Each cell has faces across which the three can flow (Fig. 5.2).

5.3 TYPES OF RESERVOIR MODEL

5.3.1 Grid Types

The reservoir division into cells can be done in terms of **Cartesian** or **radial coordinates**. These are sometimes known as **structure grids** (Fig. 5.3).

Other types are **unstructured** or **irregular grids known as PEBI grids**.

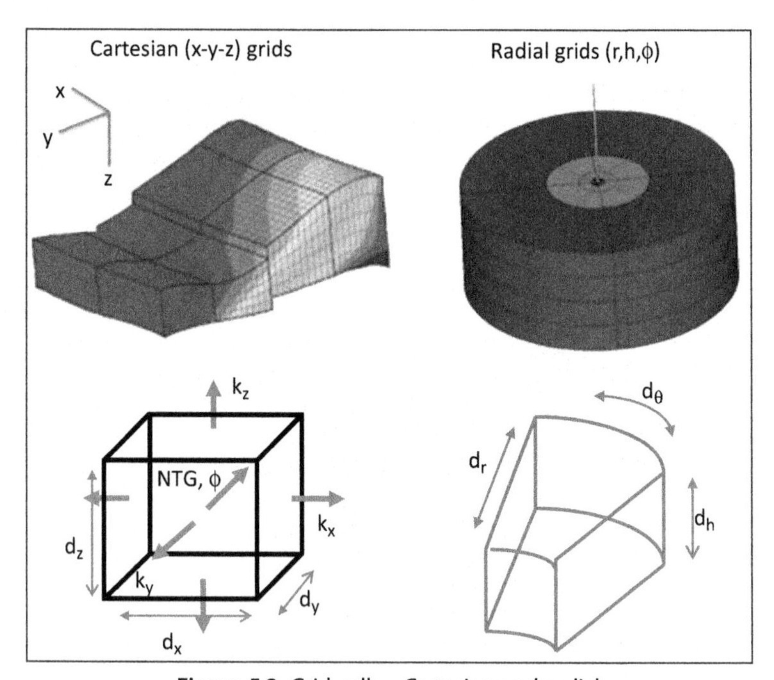

Figure 5.3 Grid cells—Cartesian and radial.

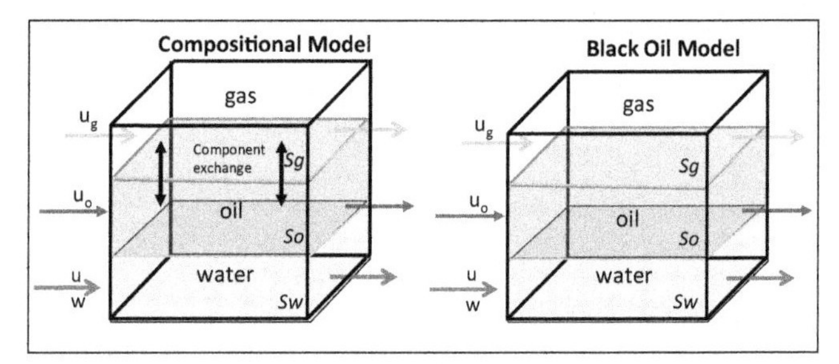

Figure 5.4 Compositional and black-oil transport schematic.

For numerical reservoir flow simulation, these grids can be made to conform to major flow features (such as faults and wells). Voronoi tessellations are obtained by various optimization methods.

5.3.2 Flow Types

There are two flow type models (Fig. 5.4).

- *Compositional models.* Here we follow the movement of separate components (ie. methane, ethane, etc.). There is phase exchange of components, and PVT properties are calculated from an equation of state and equality of chemical potentials.
- *Black-oil models.* Here we follow the movement of the phases (gas, oil, and water) only, and assume that there is no movement of components between phases. Oil and gas FVFs are from tabular input as functions of pressure. Tabular solution gas–oil ratio (GOR) is also input as a function of pressure.

5.4 BASIC EQUATIONS

5.4.1 Conservation of Mass

$$\nabla \cdot (\rho u) + Q_{\text{well}} = -\frac{\partial(\varphi\rho)}{\partial t} \qquad [5.1]$$

this can be written as

$$\frac{\partial \rho u_x}{\partial x} + \frac{\partial \rho u_y}{\partial y} + \frac{\partial \rho u_z}{\partial z} + Q_{\text{well}} = -\frac{\partial(\varphi\rho)}{\partial t} \qquad [5.2]$$

also known as the continuity equation, where ρ = density, φ = porosity, u = velocity, and Q_{well} = flow in or out of cell from a well.

It simply means that the rate of mass **flow into a volume** element across boundary − rate of mass flow **out across boundary** + any **flow in or out of a well** = rate of **accumulation of mass** within element.

5.4.2 Conservation of Momentum

From chapter "Basic Rock and Fluid Properties",

$$\nabla \overline{p}^V + \nabla \cdot \overline{\sigma}^V = \frac{1}{V_f} \int_{A_{fs}} \Psi_S \cdot dA + \int_V \rho F \; dV \tag{5.3}$$

This equation simply represents **a balance of average forces within a volume element** (due to pressure gradients and gravity), arising during steady-state flow through porous material.

With various simplifying assumptions, this equation can be reduced to Darcy's law:

$$\nabla p = -\frac{\mu}{k} u + \rho g \nabla z \tag{5.4}$$

5.4.3 Thermodynamic Relationships

A system may undergo a spontaneous change for one or both of two reasons.

1. To minimize energy.
2. To maximize entropy.

For a hydrocarbon mixture these relationships will determine the phase split of the components and the PVT behavior of these phases (gas and oil). Input of these relationships to reservoir models will be either from a reservoir fluid—matched equation of state (see chapter: Basic Rock and Fluid Properties) or via black-oil PVT tables.

5.4.4 Combined Equations—Diffusivity Equations

If we combine the conservation of mass and Darcy equations for oil, water, and gas we get the following equations.

5.4.4.1 Black-Oil Models

For the oil phase,

$$\nabla \cdot \left[\frac{k k_{ro}}{\mu_o \cdot B_o} (\nabla p_o - \gamma_o \nabla z) \right] + Q_o = \phi \frac{\partial \left(\frac{S_o}{B_o} \right)}{\partial t} \tag{5.5}$$

For the water phase,

$$\nabla \cdot \left[\frac{k k_{rw}}{\mu_w \cdot B_w} (\nabla p_w - \gamma_w \nabla z) \right] + Q_w = \phi \frac{\partial \left(\frac{S_w}{B_w} \right)}{\partial t} \tag{5.6}$$

For the gas phase,

$$\nabla\cdot\left[\frac{kk_{\text{rg}}}{\mu_{\text{g}}\cdot B_{\text{g}}}\left(\nabla p_{\text{g}}-\gamma_{\text{g}}\nabla z\right)\right]+\nabla\cdot\left[\frac{kk_{\text{ro}}\cdot R_{\text{s}}}{\mu_{\text{o}}\cdot B_{\text{o}}}\left(\nabla p_{\text{o}}-\gamma_{\text{o}}\nabla z\right)\right]+R_{\text{s}}Q_{\text{o}}+Q_{\text{g}}$$

$$=\phi\frac{\partial\left(\frac{S_{\text{g}}}{B_{\text{g}}}\right)}{\partial t}+\phi\frac{\partial\left(\frac{S_{\text{o}}\cdot R_{\text{s}}}{B_{\text{o}}}\right)}{\partial t}$$

Gas phase – includes transport of gas dissolved in oil phase [5.7]

where k = absolute permeability; k_{ro}, k_{rw}, and k_{rg} = phase permeabilities; γ_{o}, γ_{w}, and U_{o} = phase densities; S_{w} and S_{o} = phase saturations; B_{w}, B_{o}, and B_{g} = FVFs; and R_s = solution GOR.

5.4.4.2 Compositional Models

For component i, material balance (with the component able to flow through all phases k)

$$\sum_{k}^{n}\rho^{k}x_{i}^{k}\nabla\cdot u_{k}+Q_{i}=\phi\frac{\partial}{\partial t}\left(\sum_{k}^{n}\rho^{k}x_{i}^{k}S^{k}\right) \quad [5.8]$$

Substituting Darcy's equation for flow rate of phase k

$$\sum_{k}^{n}\rho^{k}x_{i}^{k}\nabla\cdot\left[\frac{kk_{\text{r}}k}{\mu_{k}}\left(\nabla p-\gamma k\nabla z\right)\right]+Q_{i}=\phi\frac{\partial}{\partial t}\left(\sum_{k}^{n}\rho^{k}x_{i}^{k}S^{k}\right) \quad [5.9]$$

where phases are $k = 1$ to $k = n$; ρ^k = molar density of phase k; x_i^k = mole fraction of component i in phase k; μ_k = viscosity of phase k; Q_i = source or sink term (well) for component i; and S^k = saturation of phase k.

The initial term represents the movement of phases in and out of the grid cell across the cell faces; the second term the flow in or out of any well in the cell; and the third term changes due to accumulation of mass in the cell.

These equations are used directly in black-oil models. For compositional models all PVT properties and compositions are determined from equation-of-state relationships at cell pressures. Calculated cell phase compositions thus allow for transfer of components between phases within the cell.

We need to solve the above set of equations for each grid cell in the model.

For a given cell i, they are in the general form

$$F_i + Q_i = \Delta M_i/\Delta t \quad [5.10]$$

where F_i = flow in and out of cell i from neighboring cells during time Δt; Q_i = flow into or out of cell i from a well; and M_i = volume change in cell i (in terms of surface volumes) in time Δt.

This set of equations must be solved simultaneously for all three phases and all cells in the model.

The above diffusivity equations cannot be solved analytically except in the simplest cases. Numerical solutions using finite difference methods are the most commonly used method with numerical reservoir simulators.

5.5 FINITE DIFFERENCES

5.5.1 Taylor Series

In mathematics, a **Taylor series** is a representation of a function as an infinite sum of terms that are calculated from the values of the function's derivatives at a single point:

$$f(x_o + \Delta x) = f(x_o) + \frac{\Delta x}{1!} f'(x_o) + \frac{\Delta x^2}{2!} f''(x_o) + \frac{\Delta x^3}{3!} f'''(x_o) + \cdots \quad [5.11]$$

where,

$$f' = \frac{\partial f}{\partial x} \quad f'' = \frac{\partial^2 f}{\partial x^2} \text{ etc}$$

So when we have a function value at some point x_o and want a value at a point $(x_o + \Delta x)$, we can approximate this using the Taylor series when we have the derivatives (gradients) at x_o.

Fig. 5.5 is an example of how as we add terms in the Taylor series we get closer to the true function ($f = \exp(x)$ in this example).

We can use this series to get approximate numerical solutions to our diffusivity equation.

If Δx is sufficiently small, we can approximate the first and second derivatives using the Taylor series equation.

5.5.1.1 For the First Derivative

If we can neglect second-order terms or higher, we get

$$f'(x_o) = [f(x_o + \Delta x) - f(x_o)]/\Delta x \quad [5.12]$$

called the "forward difference" equation.

We can also show that

$$f'(x_o) = [f(x_o) - f(x_o - \Delta x)]/\Delta x \quad [5.13]$$

called the "backward difference" equation.

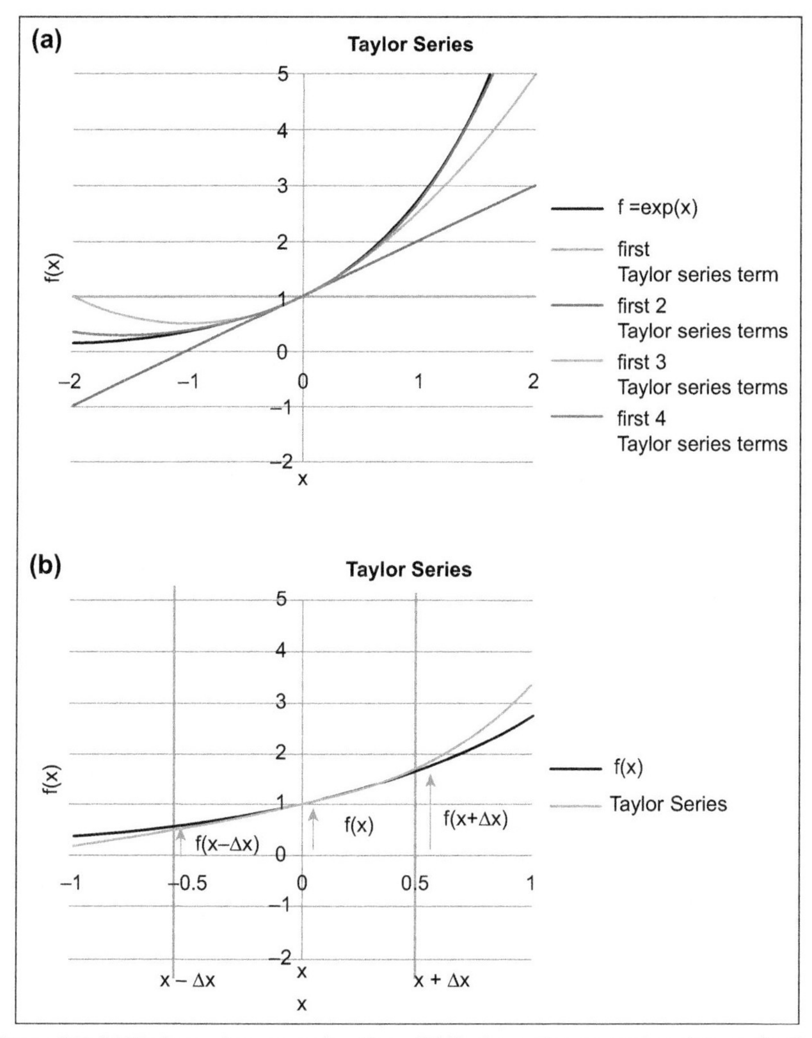

Figure 5.5 (a) Taylor series approximations. (b) Taylor series approximation neglecting second order and above.

An *average gradient* of these two would be

$$f'(x_o) = [f(x_o + \Delta x) - f(x_o - \Delta x)]/2\Delta x \qquad [5.14]$$

called the "central difference" approximation.

A useful shorthand nomenclature is

$$f_i' = (f_{i+1} - f_{i-1})/2\Delta x \qquad [5.15]$$

5.5.1.2 For the Second Derivative

Taking again forward and backward versions of the Taylor series and adding them, we get

$$f''(x_o) = [f(x_o - \Delta x) - 2f(x_o) + f(x_o + \Delta x)]/\Delta x^2 \quad [5.16]$$

or,

$$f_i'' = (f_{i-1} - 2f_i + f_{i+1})/\Delta x^2 \quad [5.17]$$

If $f = f(x,t)$ distance and time, we can write at some time step n

$$f_i'^n = \left(f_{i+1}^n - f_{i-1}^n\right)/2\Delta x \quad [5.18]$$

and,

$$f_i''^n = \left(f_{i-1}^n - 2f_i^n + f_{i+1}^n\right)/\Delta x^2 \quad [5.19]$$

5.5.2 Explicit Methods

Suppose we now look at the single-phase flow of a slightly compressible fluid in a one-dimensional, homogeneous, horizontal system. The diffusivity equation is

$$\frac{\partial^2 p}{\partial x^2} = \frac{\varphi \mu c}{k}\frac{\partial p}{\partial t} \quad [5.20]$$

which can be approximated by

$$\left(p_{i-1}^n - 2p_i^n + p_{i+1}^n\right)/\Delta x^2 = \frac{\varphi \mu c}{2k}\left(p_i^{n+1} - p_i^n\right)/\Delta t \quad [5.21]$$

or rearranging

$$p_i^{n+1} = \frac{2k}{\varphi \mu c}\cdot\left(p_{i-1}^n - 2p_i^n + p_{i+1}^n\right)\cdot\frac{\Delta t}{\Delta x^2} + p_i^n \quad [5.22]$$

We can thus estimate pressure at time $(n+1)$ from known values of pressure at time n. At the start we know p_{i-1}^n and p_{i+1}^n from initial equal reservoir pressure. This is a fully **explicit method** (Fig. 5.6). Pressure at time $t + \Delta t$ is a function of known pressures at time t.

This is applied across the grid and for each time step. If $\frac{\Delta t}{\Delta x^2}$ is too large, we can get an instability problem with oscillating solutions.

5.5.3 Implicit Methods

To overcome this instability problem, Crank and Nicolson proposed replacing $\left(p_{i-1}^n - 2p_i^n + p_{i+1}^n\right)$ by an *average value* between the $n+1$ and nth time steps.

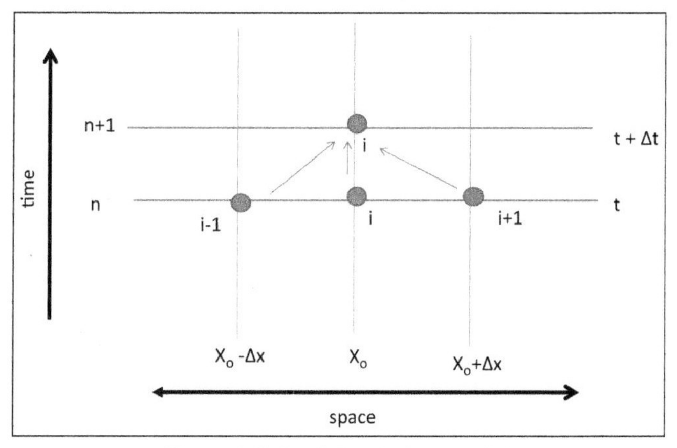

Figure 5.6 Explicit solution model.

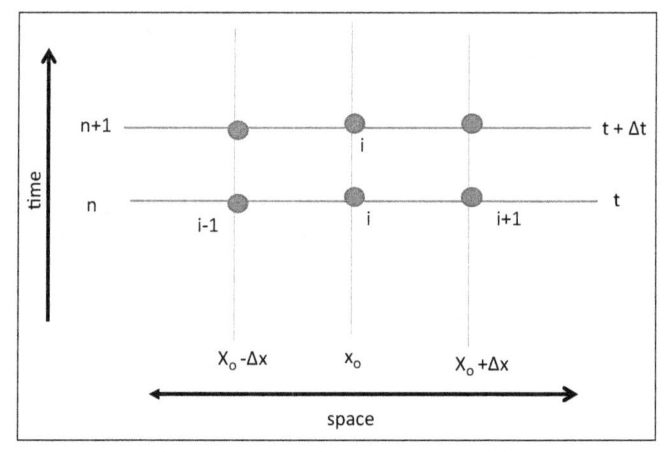

Figure 5.7 Implicit solution model.

Since a single value of the dependent variable cannot be computed explicitly for time $t + \Delta t$, this is called an **implicit method** (Fig. 5.7). Matrix algebra is employed to solve the problem.

5.6 INPUT DATA FOR NUMERICAL SIMULATORS

5.6.1 Grid Properties

These are used to define the depth and size/shape of each of the grid cells (see Fig. 5.3).

5.6.2 Rock Properties

Rock properties are the following.

1. Porosity (φ).
2. NTG.
3. Rock compressibility (c_R).

   Permeability will depend on grid type.

   For Cartesian grids

   - permeability (k_x, k_y, k_z)
   - k_v/k_h (vertical/horizontal permeability) can be specifically input.

   For radial grids:

   - permeability (k_r, k_Θ, k_h).

Input can be very simple, with porosity and permeability input manually for each grid cell, or can be generated by software **like Petrel** which uses geological input to assign grid cell properties for input to the simulator (Fig. 5.8).

5.6.3 Fluid Properties

5.6.3.1 Black-Oil Models

Tables of oil and gas properties are input to the simulator as **functions of pressure**.

1. Oil FVF (B_o).
2. Gas FVF (B_g).

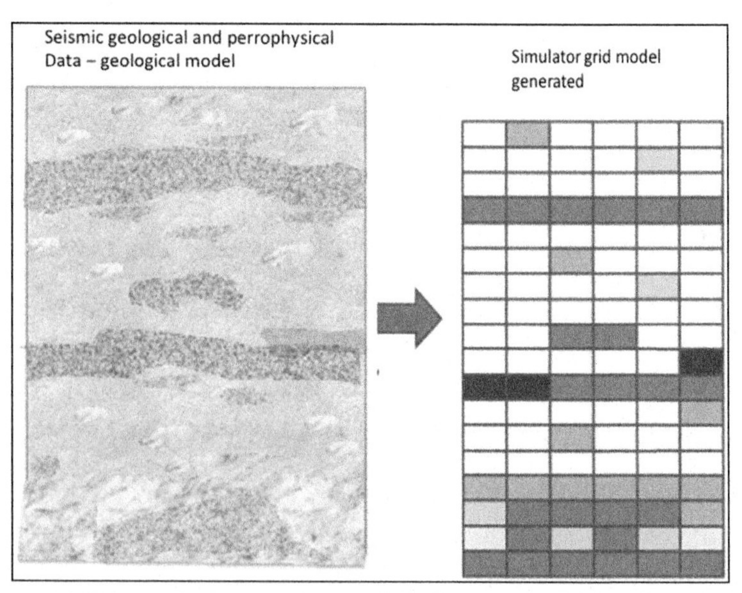

Figure 5.8 Schematic of generation of grid cell properties from geological data.

Table 5.1 Some PVT parameters

P (psi)	B_o (RB/stb)	B_g (RB/mscf)	R_s (scf/stb)	μ_o (cP)	μ_g (cP)
2000	1.467		838.5	0.3201	
1800	1.472		838.5	0.3114	
1700	1.475		838.5	0.3071	
1640	1.463	1.920	816.1	0.3123	0.0157
1600	1.453	1.977	798.4	0.3169	0.0155
1400	1.408	2.308	713.4	0.3407	0.0140
1200	1.359	2.730	621.0	0.3714	0.0138
1000	1.322	3.328	548.0	0.3973	0.0132
800	1.278	4.163	464.0	0.4329	0.0126
600	1.237	4.471	383.9	0.4712	0.0121
400	1.194	7.786	297.4	0.5189	0.0116
200	1.141	13.331	190.9	0.5893	0.0106

3. Solution GOR (R_s).
4. Gas viscosity (μ_g).
5. Oil viscosity (μ_o).

Table 5.1 is an example of such pressure black-oil tables.

5.6.3.2 Compositional Models

For compositional models we input tables of critical properties and interaction coefficients. These are obtained for the real reservoir mixture by matching an equation of state to laboratory PVT data. There are a number of commercial packages that do this very effectively. The first step is to "pseudoise" the real component mixture, where laboratory data will report perhaps 20–30 component mole fractions with a C30 + component mole fraction. Computer time limitations mean that these data need to be "lumped" into a smaller number of pseudo-components—typically 6 to 12. Starting values for critical properties and binary interaction coefficients for the pseudocomponents can be estimated from the lumped molecular weights. Regression packages will then fit these properties to laboratory data, such as constant volume depletion and constant composition expansion of differential liberation data.

Tables 5.2 and 5.3 give an example of pseudo-component equation-of-state inputs.

5.6.3.3 Dual-Porosity Models

Where we have a fractured reservoir, the model is refined to allow for the main transport to be via the fractures. We then have two cells for each

Table 5.2 Example of component critical properties

	H_2S	CO_2	PC1	C2	PC2	PC3	PC4	PC5
Tcrit (K)	373.5	304.2	189.3	305.4	390.7	546.1	749.2	924.2
Pcrit (bars)	90.010	73.8	45.9	48.8	40.4	30.0	18.3	10.5
Accentric factor	0.10	0.225	0.113	0.098	0.170	0.300	0.540	0.954
Mol wt	34.07	44.01	16.14	30.07	49.90	99.00	259.0	486.0

location, one representing the matrix and one the fractures. A further parameter is introduced determining the exchange of phases between the two.

5.6.4 Saturation Properties

Saturation properties cover relative permeability and capillary pressure. These are normally input together as functions of saturation. Table 5.4 shows an example for a two-phase oil—water system.

Where we have a three-phase system, relative permeabilities are input in a number of ways specific to the software being used.

5.6.5 Initial Reservoir Conditions

Initial reservoir conditions need to be defined: these include pressure at a given depth, depth of oil—water and gas—oil contacts. Numerical simulators have an "equilibration" keyword that enables this information to be input.

With compositional models the molar composition at a given depth must be input.

5.6.6 Well Location and Rate Control

The proposed well locations and production control, facilities constraints, and where applicable historical rates will need to be input.

1. Well locations and well controls.
2. Completion data and history.

Table 5.3 Example of binary interaction coefficients

	H_2S	CO_2	PC1	C2	PC2	PC3	PC4	PC5
H_2S	0.00							
CO_2	0.00	0.00						
PC1	0.05	0.10	0.00					
C2	0.05	0.10	0.12	0.00				
PC2	0.05	0.10	0.16	0.10	0.00			
PC3	0.05	0.10	0.03	0.10	0.10	0.00		
PC4	0.05	0.10	0.08	0.10	0.10	0.001	0.00	
PC5	0.05	0.10	0.096	0.10	0.10	0.002	0.001	0.00

Table 5.4 Example saturation properties

S_w	K_{rw}	k_{ro}	P_{cow}
0.20	0.000	0.800	2.4
0.25	0.002	0.610	1.1
0.30	0.009	0.470	0.8
0.35	0.020	0.370	0.6
0.40	0.033	0.285	0.5
0.45	0.051	0.220	0.4
0.50	0.075	0.163	0.3
0.55	0.100	0.120	0.2
0.60	0.132	0.081	0.15
0.65	0.170	0.050	0.1
0.70	0.208	0.027	0.05
0.75	0.251	0.010	0.01
0.80	0.300	0.000	0.00

3. Rate history.
4. Production controls.
5. Injection controls.
6. Facilities constraints, flowing pressures/rates.
7. Well intervention and workover strategies.

5.6.7 Aquifers

The size and strength of the underlying aquifer are often of critical importance in the long-term behavior of the field. There are a number of possible ways of modeling this. The grid system of the model can be extended to cover a significant amount of the aquifer. This is, however, often not efficient in use of computer time, and an alternative is to have a separate large aquifer grid block (like a large tank of water) connected to a number of the water-filled grid cells in the main model. The size of this "aquifer" cell and its permeability can be defined to reflect our geological understanding of the aquifer. There are also a number of "numerical" aquifer options available in the various simulators.

5.7 USE OF NUMERICAL SIMULATORS

5.7.1 Introduction

Numerical simulators are very powerful and valuable tools in planning field development of hydrocarbon resources and understanding reservoir behavior. They are, however, often misused. One particular problem is the early construction of very large models, often with hundreds of thousands or even millions of grid cells based on very limited field data. These overly complex models can be very misleading. Development plans, forecasts, and financial commitments are often made on the basis of such models and are difficult to change later. It is the aim of this section to outline efficient and effective ways in which simulators can be used to avoid this.

5.7.2 Single-Well Modeling

Numerical simulators with radial models (Fig 5.9) can be used for single-well modeling as a step forward from the single-cell models discussed above. Again, they are combined with aggregation modeling, as in chapter "Analytical Methods for Prediction of Reservoir Performance". In practice there are often convergence problems encountered, and results are little different from those obtained from the single-well spreadsheet models "gas decline-zz" and "oil imbalance."

5.7.3 Coarse Grid Modeling

While we still have limited data, coarse grid models (with only a few thousand grid cells) can be very valuable and used alongside single-well radial and aggregation models. They can be run rapidly to determine the critical factors and start to understand the basic dynamics of the reservoir. An example is shown in Fig. 5.10.

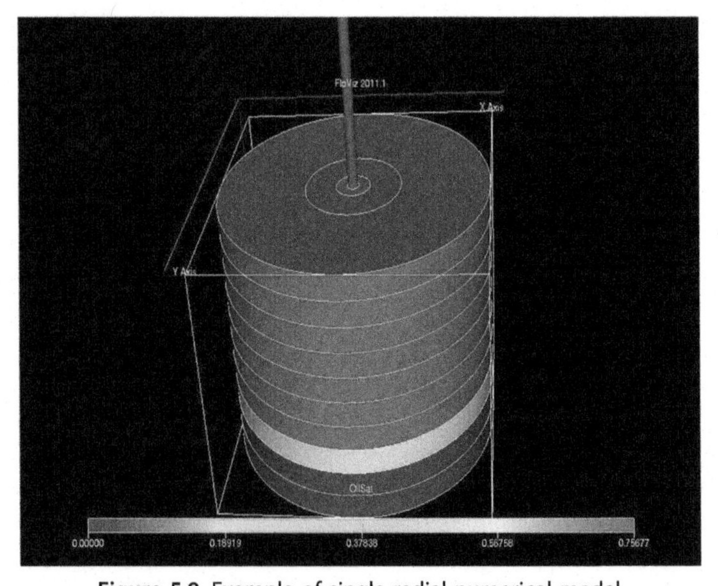

Figure 5.9 Example of single radial numerical model.

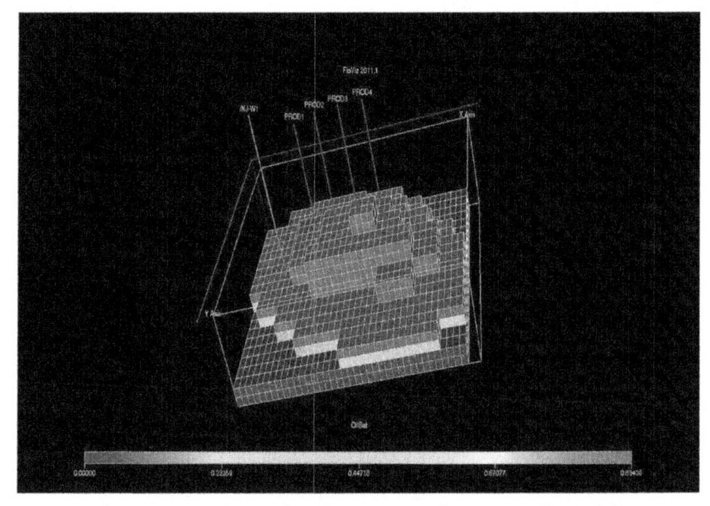

Figure 5.10 Example of coarse grid numerical model.

5.7.4 Conceptual/Sector Modeling

5.7.4.1 General

Initial reservoir modeling should always be aimed at understanding the fundamental dynamics of the field. A sector model should be built. This will cover a representative part of the field with the aim of finding out the critical parameters that determine achievable production, recovery, and economics.

For example, in a planned five-spot water-flood development a sector model like that shown in Fig. 5.11 might be constructed.

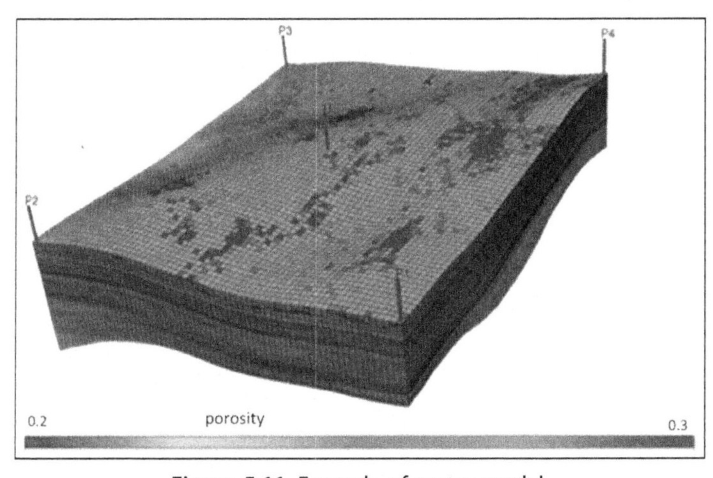

Figure 5.11 Example of sector model.

5.7.4.2 Sensitivity Analysis

A sensitivity analysis can then be carried out to determine the parameters, with their range of uncertainty that will have the largest effect on recoverable reserves and net present value (NPV). A base-case model with best estimate values for all parameters is run.

Then for each parameter, with all others held constant, the model is run with the upper and lower values of this parameter. The results enable a "tornado diagram," which is discussed in chapter "Field Appraisal and Development Planning".

The most important parameters that should be considered will depend on the type of reservoir, but all typical uncertainties with a significant effect on reservoir performance are shown below.

1. Volumetric parameters
   a. gross rock volume
   b. porosity/NTG.
2. Recovery factor (RF) parameters
   a. permeability
   b. k_v/k_h
   c. aquifer strength
   d. heterogeneity
   e. strata permeability ratio (potential for stronger water advance)
   f. relative permeability and in particular residual saturations
   g. GOR and its dependence on depth can be important for gas condensate reservoirs
   h. in gas–water systems the base and top geological structure can be critical.

Results from this analysis are used for two purposes: further appraisal planning and field development planning. From our tornado diagram the top two or three critical parameters are selected—all the rest can be, at least provisionally, ignored.

5.7.4.3 Appraisal Planning—Value of Information

The use of value of information in planning field appraisal is discussed in chapter "Field Appraisal and Development Planning". We need to look at what data acquisition is necessary to narrow the uncertainty on the two or three parameters identified in the sensitivity analysis. This is best carried out with a simple numerical sector model.

5.7.4.4 Field Development Planning

The first stage in field development planning is to use our base-case gross rock volume (GRV, upscaled to full reservoir size), petrophysical parameter average and RFs, as discussed in Section 5.2, to optimize an economic development scheme. We can then use a simple Monte Carlo analysis with the ranges of our three basic parameters—GRV, petrophysics, and RFs (with the major sensitivity range only)—to determine upside (P10) and downside (P90) cases and look at economic sensitivity.

The next stage is to use what is known as "experimental design" methods with the reservoir sector model, which gives a more sophisticated answer allowing for the interaction of parameter effects more fully. This topic is covered briefly in Chapter 11 (Section 11.3.3.2).

5.7.5 Full-Field Modeling

There are situations where full-field modeling may be needed at an early stage. This will be where structural uncertainties are critical (eg, case (g) above). In this case cell numbers should be limited and overcomplexity avoided. The general approach outlined above for sector modeling is then applied to a simple model of the whole field.

At a later stage, when we have a significant amount of historical production and other data, the construction of larger and more complex full-field models is then justified.

5.8 HISTORY MATCHING

History matching is the process by which a reservoir model is adjusted to match the production and pressure history of the reservoir. A history-matched reservoir simulation model will more accurately predict the future performance and better represent the current pressure and saturation state of the reservoir.

5.8.1 What Is History Matched?

Production data are fluid rates (oil, water, and gas), fluid components, and tracers. Pressure data are repeat formation testing, bottom hole pressure (BHP), tubing head pressure (THP), and continuous downhole monitoring.

Saturation distribution data are for the well and 4D seismic.

Pressure is the most commonly matched value, followed by water cut. A common problem is a lack of BHP data due to the absence of permanent downhole gauges. THP must be used, along with some estimate of wellbore pressure drop.

5.8.2 What Is Changed to Achieve a History Match?

The main factors changed in achieving a history match are as follows:

Permeability (especially distribution).

Porosity (or other factors affecting pore volume).

Initial fluid distribution.

PVT, relative permeability, capillary. pressure, rock compressibility, etc.

Faults (transmissibility, location).

Wells (completions, productivity index (PI)).

A very common error in matching pressures where we have a large field with many wells is the local adjustment of permeabilities and porosities around wells so that we have a patchwork "sticking-plaster" approach. It is possible to match almost anything this way, but it tells us nothing about the reservoir as a whole. It is much better to try to find global (or at least regional) parameter changes that improve the history match and our understanding of what is going on in the field.

5.9 QUESTIONS AND EXERCISES

Q5.1. Explain the different types of numerical reservoir simulation models that can be used to predict reservoir behavior.

Q5.2. Outline reservoir model construction with reference to the different sections of typical simulator input, starting from the selection of the type of models available. Describe the difference in input required for compositional and black-oil models.

Q5.3. Discuss history matching with numerical simulators. What is typically "matched", and what parameters are used to achieve a match?

Q5.4. The following three diffusion equations (for oil, water, and gas) are the basis of numerical black-oil reservoir models:

$$\nabla \cdot \left[\frac{kk_{\mathrm{ro}}}{\mu_{\mathrm{o}} \cdot B_{\mathrm{o}}} (\nabla p_{\mathrm{o}} - \gamma_{\mathrm{o}} \nabla z) \right] + Q_{\mathrm{o}} = \phi \frac{\partial \left(\frac{S_{\mathrm{o}}}{B_{\mathrm{o}}} \right)}{\partial t}$$

$$\nabla \cdot \left[\frac{kk_{\mathrm{rw}}}{\mu_{\mathrm{w}} \cdot B_{\mathrm{w}}} (\nabla p_{\mathrm{w}} - \gamma_{\mathrm{w}} \nabla z) \right] + Q_{\mathrm{w}} = \phi \frac{\partial \left(\frac{S_{\mathrm{w}}}{B_{\mathrm{w}}} \right)}{\partial t}$$

$$\nabla \cdot \left[\frac{kk_{rg}}{\mu_g \cdot B_g}(\nabla p_g - \gamma_g \nabla z)\right] + \nabla \cdot \left[\frac{kk_{ro} \cdot R_s}{\mu_o \cdot B_o}(\nabla p_o - \gamma_o \nabla z)\right]$$

$$+ R_s Q_o + Q_g = \phi \frac{\partial\left(\frac{S_g}{B_g}\right)}{\partial t} + \phi \frac{\partial\left(\frac{S_o \cdot R_s}{B_o}\right)}{\partial t}$$

where $\gamma = \rho g$.

Write equations for the conservation of mass and Darcy's law.

Explain in terms of these equations the significance of the various parts of the oil-phase diffusion equation shown above.

Q5.5. Write the Taylor series up to the third derivative term. Explain with the aid of diagrams the difference between explicit and implicit methods for solving diffusion equations.

5.10 FURTHER READING

J.R. Fanchi, Principles of Applied Reservoir Simulation, 2006.

K. Aziz, A. Settari, Petroleum Reservoir Simulation, Elsevier, 1979.

CHAPTER 6

Estimation of Reserves and Drive Mechanisms

This chapter covers early estimation of hydrocarbons initially in place and the recovery factors (RFs) that we may expect for the various types of reservoir.

6.1 HYDROCARBONS IN PLACE

6.1.1 Hydrocarbon Pore Volume

Hydrocarbon pore volume is determined from the geological (area and average reservoir thickness) and petrophysical (porosity and net to gross—NTG) input (Fig. 6.1). Where we have limited data in early field life, we take

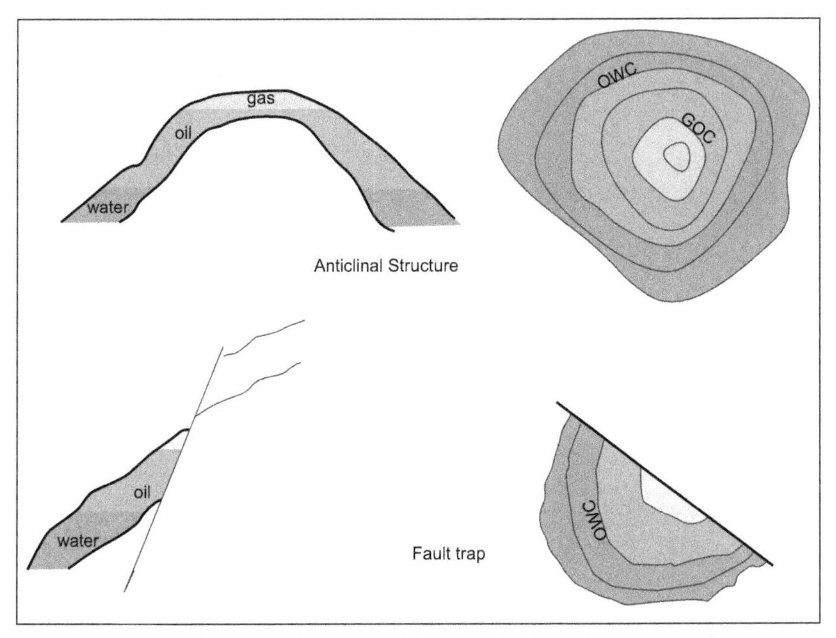

Figure 6.1 Examples of structural input to hydrocarbon pore volume.

Fundamentals of Applied Reservoir Engineering
ISBN 978-0-08-101019-8
http://dx.doi.org/10.1016/B978-0-08-101019-8.00006-5
© 2016 Elsevier Ltd.
All rights reserved.

single values for reservoir area and average values for net thickness, porosity, and water saturation, so that

$$V = Ah_v\phi(1 - S_w) \quad [6.1]$$

where A, area (average); h_v, net thickness, $h \cdot$ NTG; ϕ, porosity; and S_w, water saturation; NTG, net to gross.

When detailed appraisal data become available, the reservoir regions and areas will of course be given separate values, but the same equation will be used to calculate regional pore volumes.

6.1.2 Oil in Place

Using the above equation for hydrocarbon pore volume, stock tank oil in place is given in field units by:

$$N = 7758Ah_v\phi(1 - S_w)/B_{oi} \quad [6.2]$$

where N, stock tank oil initially in place; A, area in acres; h_v, net thickness, $h \cdot$ NTG in feet; B_{oi}, the initial oil formation volume factor in Rb/stb; and 7758, conversion factor RB/acre-ft.

6.1.3 Gas in Place

For gas in place:

$$G = 7758Ah_v\phi(1 - S_w)/B_{gi} \quad [6.3]$$

where G, gas in place in scf; A, area in acres; h_v, net thickness, $h \cdot$ NTG in feet; B_{gi}, gas formation volume factor in Rb/scf; and 7758, conversion factor RB/acre-ft.

$$B_g = \frac{P_b\, TZ(P)}{5.615P\, T_b Z_b} \quad [6.4]$$

Now standard conditions are P_b, 14.7 psi; Z_b, 1; T, °Rankin (= 60°F + 460), so that:

$$B_{gi} = 0.0283TZ/p(\text{res bbl/scf})$$

An example of variation of Z factors with pressure is shown in Figure 2.40(b).

6.2 RESERVES

Reserves are simply the oil or gas in place times the RF, so for an oil reservoir:

$$R = 7758Ah_v\phi(1 - S_w)/B_{oi} \cdot \text{RF} \quad [6.5]$$

in stb and for a gas reservoir:

$$R = 7758Ah_v\phi(1 - S_w)/B_{gi} \cdot \mathrm{RF} \qquad [6.6]$$

in scf

6.3 RECOVERY FACTORS FOR VARIOUS FIELD TYPES

The fraction of the hydrocarbons initially in place that can be recovered will depend on the effectiveness of the "drive mechanism," ie, what is driving the gas or oil toward the producing wells and how efficient it is.

We thus need to look at each hydrocarbon reservoir type separately when we consider what proportion of the hydrocarbon in place we might reasonably expect to recover economically.

6.3.1 Dry and Wet Gas Reservoirs

Gas is highly compressible, so the drive mechanism here is gas expansion. The total recoverable reserves will depend on the initial pressure, the final abandonment pressure, and the PVT properties of the gas. The simple relationship between gas produced and pressure drop ($\Delta V = f(p)$) was discussed in Chapter 4. The decline in gas rate with time ($q = \Delta V/\Delta t = f(t)$) is not so simple. It depends on the initial rate, the volume and geometry of the reservoir, the permeability distribution, and how the well is flowed. The subject was discussed in Chapter 4 when we looked at decline curves. Gas reservoirs have high RFs, typically between 65% and 95%.

An important factor in determining the RF in gas fields is the possible "watering out" of wells. This will depend on the presence of high-permeability layers that result in what are called "stringers" of water advancing ahead of the main aquifer advance as pressure decreases (see Fig. 6.2).

A similar problem is water "coning" where local pressure gradient around a producing well sucks up water from the aquifer (Fig. 6.2). For this reason wells are normally completed well above the gas/water contact.

Natural fractures can have a similar effect, reducing sweep efficiency.

Once water hits a production well, if the water cut (percentage of water in produced liquids) increases too rapidly, the well is almost certainly lost. It is particularly important therefore in estimating the RF achievable in a gas reservoir to consider the geology of the reservoir and the potential for premature water breakthrough.

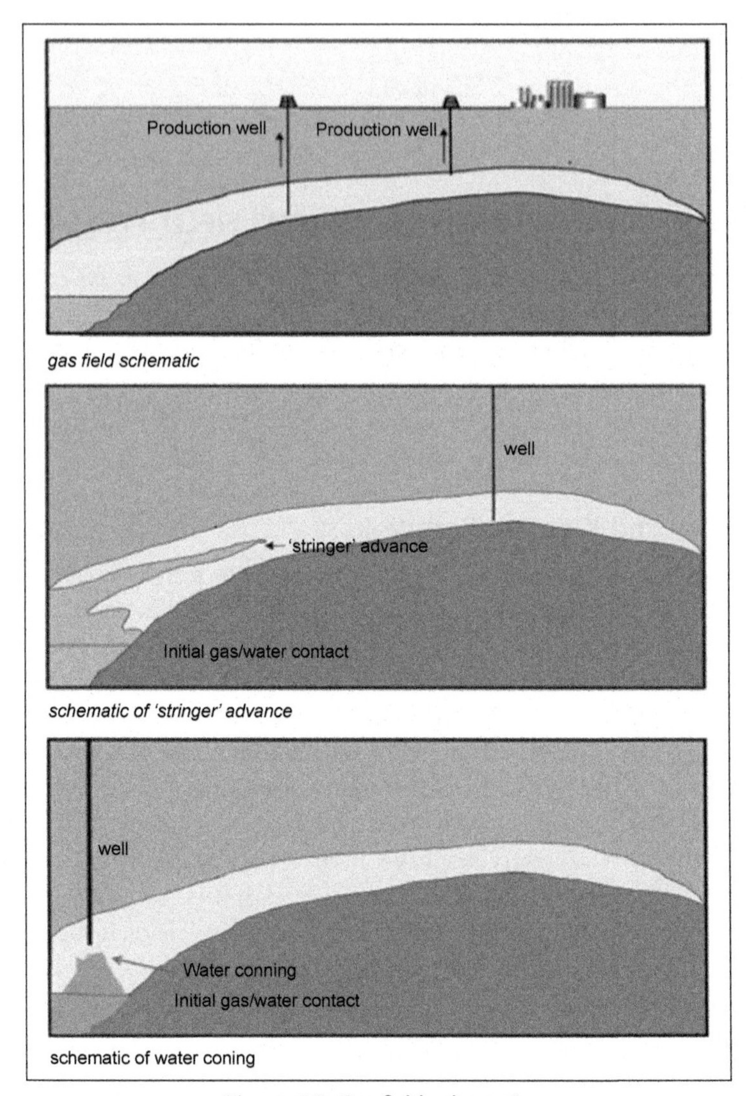

Figure 6.2 Gas field schematic.

To try to avoid these problems, wells are normally completed as high as possible in the reservoir, and use of highly deviated or horizontal wells is often considered. The size and strength of the underlying aquifer must also be examined.

6.3.2 Gas Condensate Reservoirs

As we saw in Chapter 2, the distinguishing feature of gas condensates is that liquids drop out *in the formation* once reservoir pressure goes below the dew point. There are two consequences of this.

1. Liquid dropout is normally largely immobile, so significant amounts of valuable surface liquids are lost.
2. Some of this liquid dropout around wells can partially block gas movement into the well, reducing well deliverability.

For these reasons, and because the value of the condensate is normally significantly more than the value of the gas, straight depletion of gas condensate fields would not usually be considered a reasonable development option. Gas recycling would be a normal development. Here we attempt to keep the reservoir above the dew-point pressure for as long as possible by reinjecting some or all of the dry gas following the separation process in the surface separation facilities (see Fig. 6.3). These will normally be multistage separators to maximize the heavy and middle hydrocarbon recovery (see ternary diagram in Fig. 6.3). Recycling will also tend to sweep the liquid-rich gas toward the producing wells. At some point, when further recycling is no longer considered economically efficient, we have a straight depletion process (known as blowdown).

RFs are difficult to predict for gas condensate fields, but liquid recovery of >40% can be achieved and final blowdown may result in gas recovery of 70–80%. Because liquids are the more valuable product, this liquid recovery is normally the most important figure.

It should be remembered when looking at gas condensate fields that there can be significant variation in hydrocarbon composition with depth (when compared with oil fields), with heavier components more prevalent with increasing depth. This can be a relatively small effect, but may become evident from initial fluid test samples.

6.3.3 Undersaturated Oil Fields

6.3.3.1 Liquid Expansion Drive

Expansion of the original oil between initial pressure (pi) and p (bubble point) will provide a drive for production. Since the compressibility of liquids is small, recovery from oil expansion is normally very small (<10%), and only if initial reservoir pressure is very much higher than bubble-point

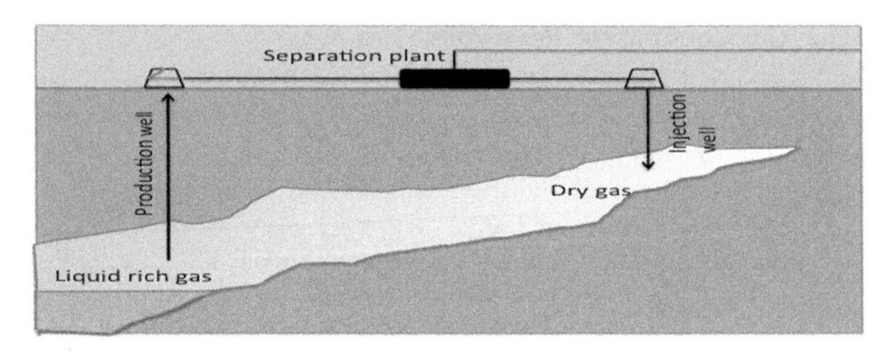

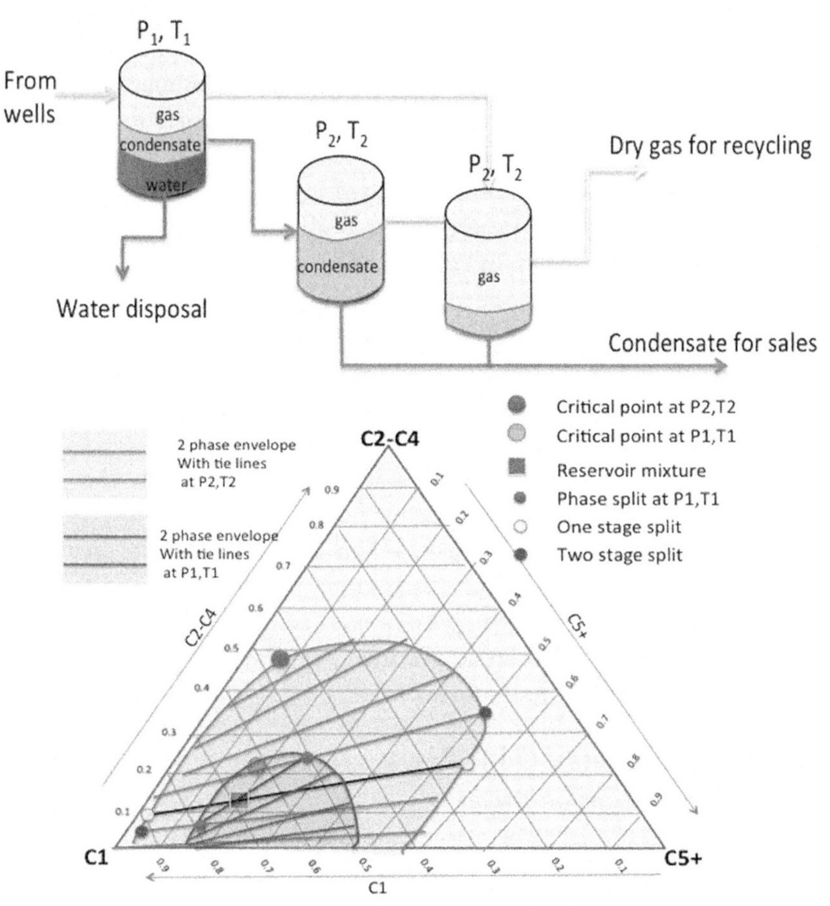

Figure 6.3 Gas condensate recycling schematic.

pressure (a somewhat unusual situation) will significant recovery due to this drive be possible.

6.3.3.2 Solution Gas Drive

There are two components here.

1. Expansion of original oil between pi and p.
2. Expansion of liberated gas, normally the major effect.

RFs from solution gas drive may typically be around 25–35%.

What happens to the released gas is important. Gas evolved around the well as pressure drops can do one of the following.

1. Remain immobile close to the well.
2. Migrate into the well and be produced.
3. Travel upwards to form a gas cap or add to an existing cap.

Migration into the well will reduce recovery (losing some of the gas expansion drive), and also disposal of unwanted gas can be a serious problem. In Fig. 6.4, if we had no loss of gas to the well the situation would be that on the left-hand side in the figure. Solution gas drive would be driven by a combination of growth of a created gas cap and expansion of gas bubbles in the oil.

6.3.3.3 Water Flooding

Water flooding is a major development method for oil reservoirs and is covered in some detail in Chapter 4. Water is injected from some wells to maintain reservoir pressure as oil is produced from others. The aim is to position injection and production wells such that we "sweep" the oil toward the producing wells.

RFs can be as high as 60%, but will depend on sweep efficiency both across the area and "locally" with respect to rock/fluid properties. Thus total sweep efficiency may be written as:

$$E_T = E_{R/F} * E_A \qquad [6.7]$$

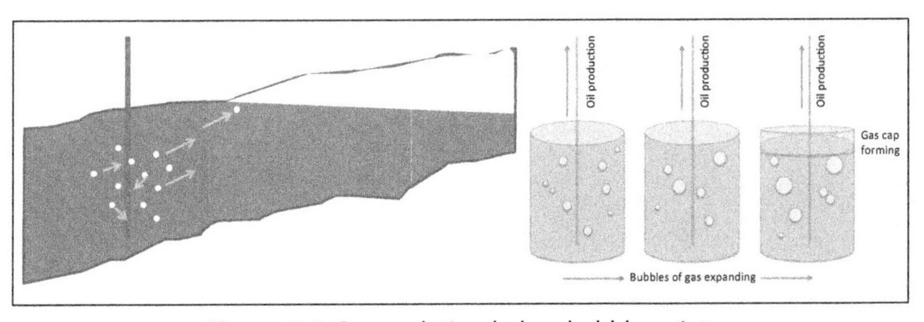

Figure 6.4 Gas evolution below bubble point.

where E_T, total sweep efficiency, E_A, areal efficiency, and $E_{R/F}$, rock/fluid-dependent sweep efficiency.

E_A depends on the extent of the contact of the advancing water front with the formation. It has horizontal and vertical components and will depend on the level and type of heterogeneity in the reservoir. For example, high-permeability layers connected between injectors and producers will reduce E_A as water preferentially flows through these layers and has a poorer sweep of the lower-permeability areas.

$E_{R/F}$ will depend on wettabilities, relative permeabilities (particularly residual oil saturation), and fluid viscosities. It assumes a locally homogeneous-type behavior. This aspect of water-flooding efficiency is covered in detail in Chapter 2.

Understanding of areal sweep efficiency only comes with detailed appraisal, analysis of production data, and detailed numerical modeling. Various well layout patterns are used to maximize recovery—some examples are shown in Fig. 6.5.

It is particularly important to maintain pressure above or close to the bubble point to prevent reduction of effective permeability of the oil and also production of unwanted gas. A further complication may be the presence of natural fractures. These can tend to channel injected water from

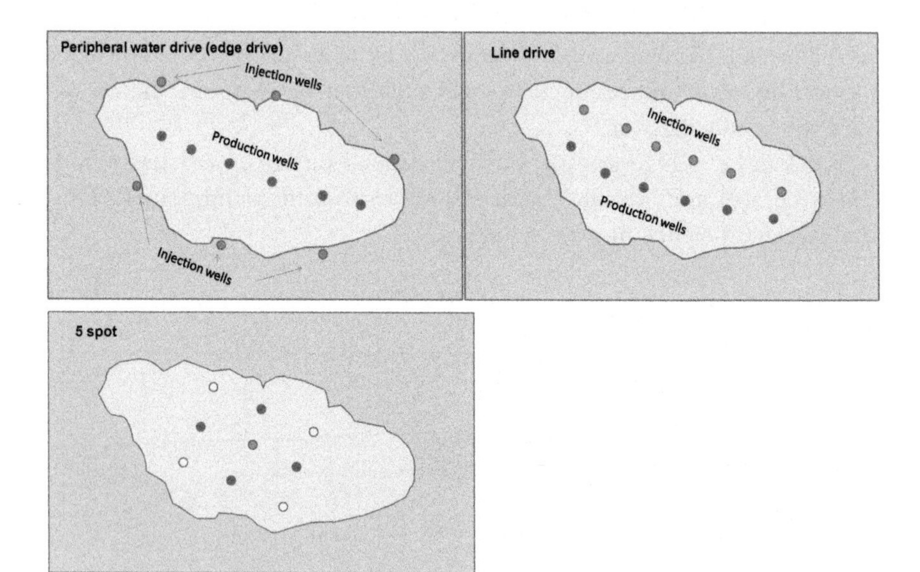

Figure 6.5 Various water-flood well layouts.

injection to production wells, bypassing significant areas of the reservoir and hence reducing sweep efficiency.

RFs for water flooding normally lie between 40% and 60%.

6.3.4 Saturated Oil Fields

These are oil reservoirs with a gas cap (Fig. 6.6). Issues are basically the same as for undersaturated oil fields, but now we have an additional drive mechanism in the expansion of the gas cap potentially driving oil toward producing wells. There is, however, the possibility of gas coning down to the wells.

The only really effective way of modeling these fields is by numerical simulation. The range of potential RFs is wide, at 20–60%. Reservoirs with thin oil layers between the gas cap and aquifer can be difficult to manage, with gas being pulled down and water coning up.

6.3.5 Enhanced Oil Recovery

This is discussed in Appendix 4.

6.3.6 Field Management

Once a field is producing, final recovery is optimized by good reservoir management. Reservoir monitoring with bottom hole, top hole, and repeat formation pressure testing and measurement of water, oil, and gas production enable detailed numerical modeling and history matching. Models

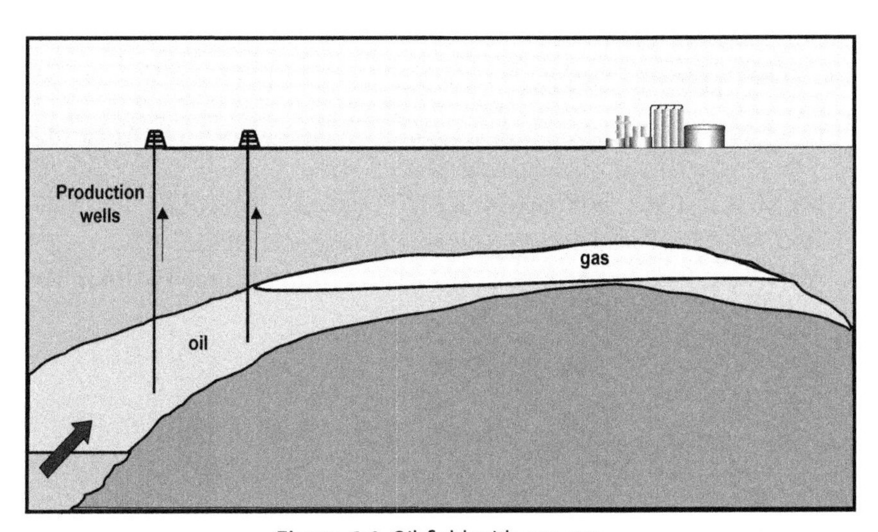

Figure 6.6 Oil field with gas cap.

are then used to optimize existing well injection and production rates, and the number and location of new wells.

6.4 QUESTIONS AND EXERCISES

Q6.1. Estimate the oil in place in million stock tank barrels in the anticlinal structure shown in Fig. 6.1 with an area of 2400 acres, assuming an NTG of 90%, porosity of 15%, and water saturation of 20%. Assume that the reservoir pressure is just below the bubble point. Use the oil formation volume factor plot shown in Fig. 2.43.

Q6.2. Estimate the gas in place in the same anticlinal structure as given in Q6.1, assuming an NTG of 90%, porosity of 15%, and water saturation of 20%. Assume that the reservoir pressure is 2000 psi and the temperature is 150°F. Use the compressibility plot shown in Fig. 2.41(b).

Q6.3. Discuss the major risks in recovery for gas fields.

Q6.4. Give typical RFs for dry gas, gas condensate, oil fields under depletion, and water-flooded oil fields. What are the drive mechanisms for each of these?

Q6.5. Explain how recovery of liquids in gas condensate fields can be increased.

Q6.6. Explain the difference between saturated and undersaturated reservoirs.

Q6.7. Under what conditions can oil expansion drive give reasonable RFs?

6.5 FURTHER READING

C. Conquist, Estimation and Classification of Reserves of Crude Oil, Natural Gas and Condensate, SPE, 2001.

M. Muskat, Physical Principles of Oil Production, McGraw-Hill, 1949.

H.B. Bradley, Petroleum Engineering Handbook, SPE, 1987.

M. Walsh, L. Lake, A Generalized Approach to Primary Hydrocarbon Recovery, Elsevier, 2003.

CHAPTER 7

Fundamentals of Petroleum Economics

7.1 INTRODUCTION

Decisions on investment in any oil or gas field development will be made on the basis of its value.

This "value" is judged by a combination of a number of economic parameters:

- net present value (NPV)
- estimated monetary value (EMV)
- real rate of return (RROR), sometimes called the internal rate of return
- profit-to-investment ratio (PI)
- payback time.

All these need to be taken into account in a full investment decision, and are considered in this chapter. Which of these indicators are considered more important than others will depend on a number of commercial and political factors, and the size and circumstances of a company.

7.2 NET CASH FLOW

Net cash flow from investment is made up of a number of components – some positive, some negative – so for example capital expenditure (CAPEX) costs of drilling wells, laying pipelines and building facilities along with operational expenditure (OPEX) must be counted against profits from selling oil or gas (Fig. 7.1). Net cash flow is normally calculated for uniform time intervals – quarterly or half-yearly.

Thus:

$$\mathrm{NCF}(i) = \mathrm{Capex}(i) + \mathrm{Opex}(i) + \mathrm{Sales}(i) \qquad [7.1]$$

where $\mathrm{NCF}(i)$ = net cash flow for period i, $\mathrm{Capex}(i)$ = capital expenditure (drilling, facilities' costs) for period i and $\mathrm{Opex}(i)$ = operating expenditure (maintenance, transportation costs) for period i.

Fundamentals of Applied Reservoir Engineering
ISBN 978-0-08-101019-8
http://dx.doi.org/10.1016/B978-0-08-101019-8.00007-7
© 2016 Elsevier Ltd.
All rights reserved.

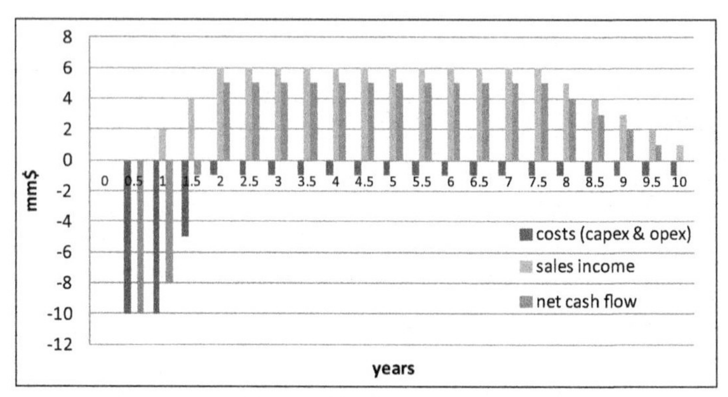

Figure 7.1 Cash flows.

7.3 INFLATION

When considering value, a number of factors must be taken into account – one of these is inflation.

Inflation is a measure of the decreasing purchasing power of money with time.

Cash flows can be described either as "nominal," by quoting the actual cash flows in each period, or as "real," by adjusting nominal cash flow for a given period to an equivalent cash flow at a fixed reference date by allowing for the cumulative effect of inflation between the reference date and the given cash flow period.

7.4 DISCOUNTED CASH FLOW

Net cash flow must be adjusted to allow for the cost of capital needed to carry out the project and develop the field.

Discount rate is either the *cost* to acquire additional capital (for example by borrowing from a bank), or the return that could be obtained by investing in an alternative opportunity (ie, if the oil company has all or part of the capital needed to develop a field, it could alternatively have invested this in some other opportunity).

Discounted cash flow (DCF) is calculated by:

$$\mathrm{DCF}_i = \mathrm{NCF}_i/(1 + r_\mathrm{D})^n \qquad [7.2]$$

where NCF = net undiscounted cash flow for period i; r_D = discount rate (fraction); and n = number of time intervals.

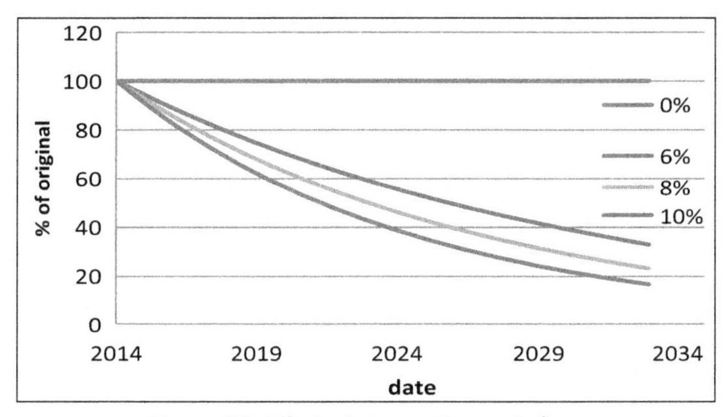

Figure 7.2 Effect of discounting cash flow.

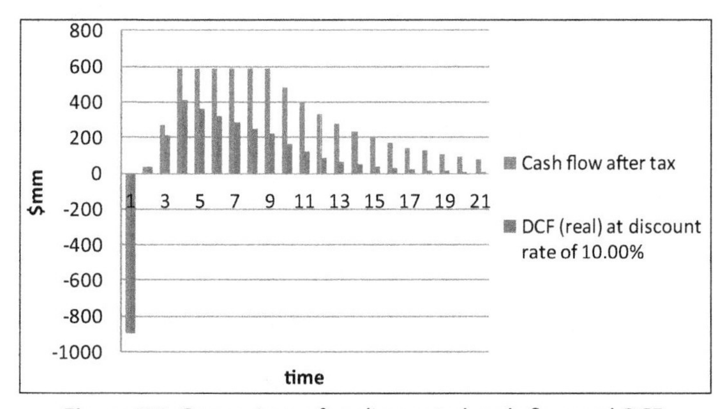

Figure 7.3 Comparison of undiscounted cash flow and DCF.

The higher the discount rate assumed, the less profitable the project will appear. Typically in the oil industry discount rates between 6 and 10 per cent are used. The effect of this is shown in Fig. 7.2.

A comparison of undiscounted cash flow and DCF is shown in Fig. 7.3.

7.5 NET PRESENT VALUE

NPV is defined as the total present value of a series of cash flows discounted at a specific rate to specific data. It is therefore a cumulative cash flow:

$$\mathrm{NPV}(r\%) = \sum_{i}^{n} \mathrm{DFC}(i) \qquad [7.3]$$

If cash flows are in real terms (allowing for inflation), a real NPV is generated. If nominal cash flows are used (not allowing for inflation), a nominal NPV is generated.

The discount rate and the nominal or real basis should always be quoted.

So we can have, for example, the following.

1. NPV10 (real), NPV0 (real).
2. NPV10 (nominal), NPV0 (nominal).

Examples of the relationship between DCF and NPV are shown in Figs. 7.4 and 7.5.

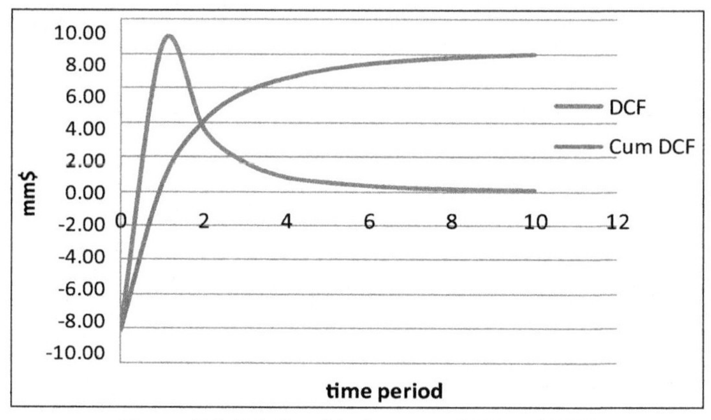

Figure 7.4 Relationship between DCF and NPV.

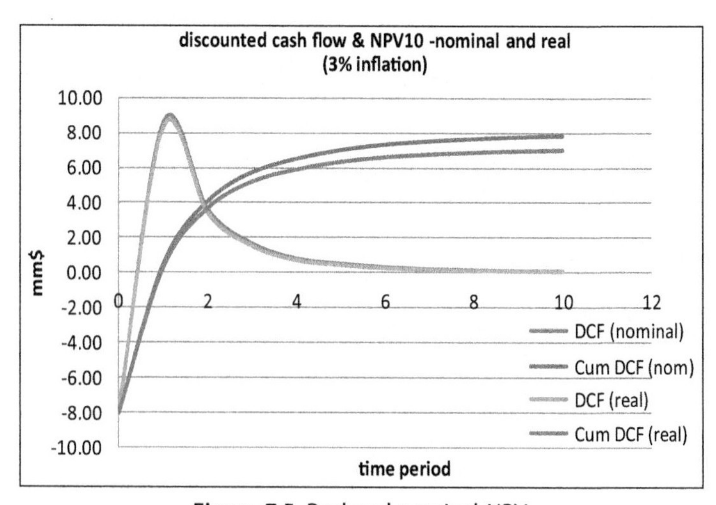

Figure 7.5 Real and nominal NPV.

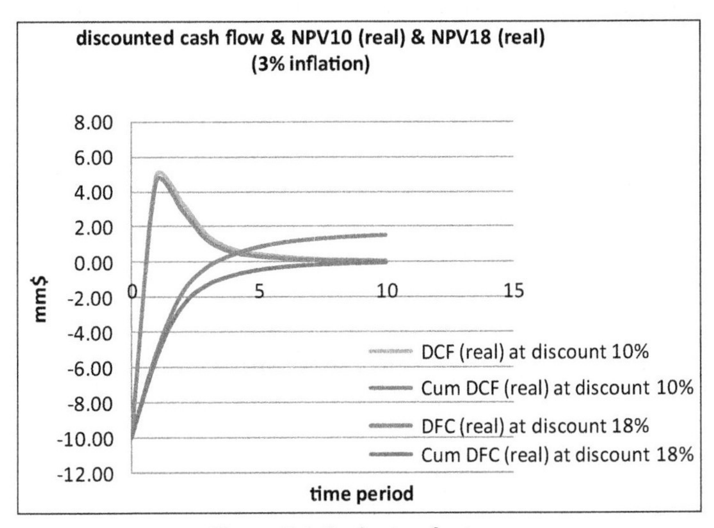

Figure 7.6 Real rate of return.

7.6 REAL RATE OF RETURN

RROR = the discount rate which must be applied to cash flow to reduce NPV real to zero. This is sometimes called the internal rate of return.

In the case shown in Fig. 7.6, RROR = 18 per cent, ie, at anything below 18 per cent discount rate the project becomes economic (NPV > 0).

7.7 PAYBACK TIME AND MAXIMUM EXPOSURE

The payback time of a project is the length of time that will elapse before the cumulative undiscounted real cash flow becomes positive (ie, before this costs outweigh income).

Maximum cash exposure is defined as the maximum negative undiscounted cumulative real net cash flow (Fig. 7.7).

7.8 PROFIT-TO-INVESTMENT RATIO

The discounted real PI is defined as:

$$PI = NPV/\text{total discounted real capital expenditure}$$

where all the discounting is done at the same rate (Fig. 7.8). This is sometimes known as return on investment, as it is a simple measure of return on investment. It is independent of time, which limits its usefulness as an economic indicator. It is always used in conjunction with other indicators.

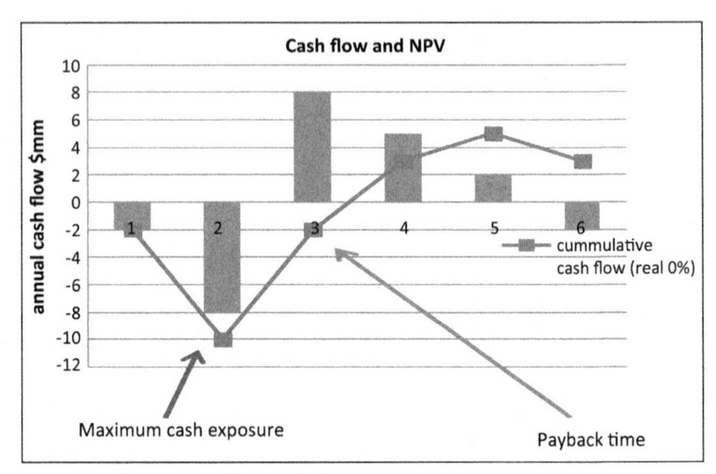

Figure 7.7 Payback time and maximum exposure.

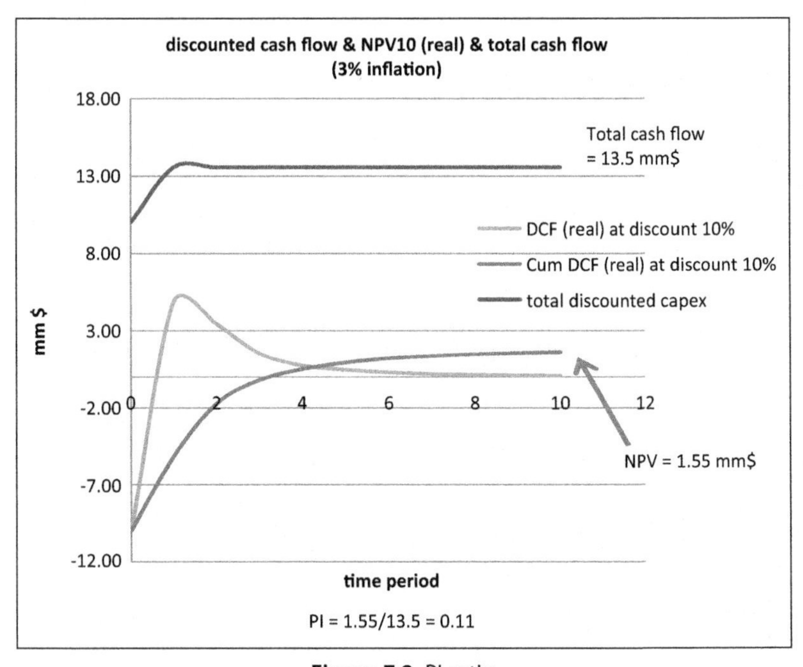

Figure 7.8 PI ratio.

7.9 RISKED INDICATORS – ESTIMATED MONETARY VALUE

All the above indicators assume a single production outcome. In reality there is always uncertainty in predicted production profiles and oil or gas prices, which should be taken into account in decision-making.

When a single "best estimate" case only is used, the indicator is known as an *unrisked indicator*. EMV is a risked indicator.

EMV is defined as:

$$\mathrm{EMV}(r) = \sum_{i}^{n} P_i \mathrm{NPV}(r_\mathrm{D})_i \quad [7.4]$$

which is the weighted sum of the NPVs corresponding to different possible outcomes, where $r =$ discount rate and P_i is the probability of outcome i. The sum of the probabilities must equal one.

A normal assumption would be that we take three outcome cases: $P_{(\mathrm{P90\ case})} = 0.25$, $P_{(\mathrm{P50\ case})} = 0.5$ and $P_{(\mathrm{P10\ case})} = 0.25$.

To take an example with downside, base-case and upside production profiles:

$$\begin{aligned}\mathrm{EMV}(10) = {} & 0.25 * \mathrm{NPV10(downside)} + 0.50 * \mathrm{NPV10(best\ est.)} \\ & + 0.25 * \mathrm{NPV10(upside)}\end{aligned} \quad [7.5]$$

We then have a project monetary value (effectively an NPV10(real)) adjusted to allow for both technical and economic uncertainty or either one alone.

The result will depend on the relationship between the three cases, which can skew the EMV up or down from the base-case NPV. Risked as well as unrisked cases should always be presented when investment decisions are to be made.

7.10 ECONOMIC INDICATOR SOFTWARE

Excel software on "economic indicators" is available for calculation of the major economic indicators NPV, PI and RROR discussed above. Gas and oil rates are input, along with drilling details, well and facilities' costs and assumed discount rate and oil and gas prices. An example (for a wet gas field) is shown Fig.7.9.

7.11 EXAMPLES WITH ECONOMIC INDICATORS

Below are a set of cases: we start from a base case, and look at the effect on indicators of a higher discount rate, higher CAPEX and a shorter plateau. These are illustrated in Figs. 7.10–7.13.

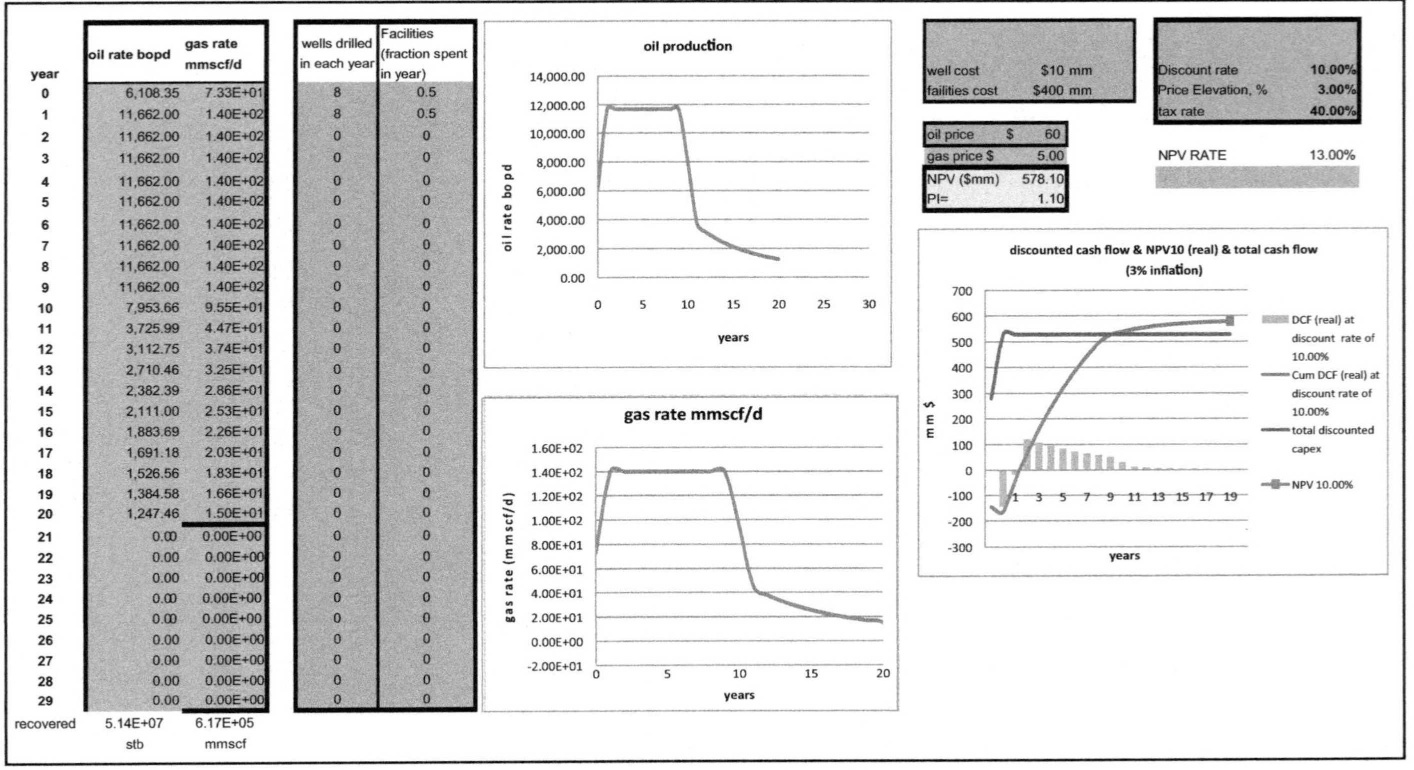

year	oil rate bopd	gas rate mmscf/d	wells drilled in each year	Facilities (fraction spent in year)
0	6,108.35	7.33E+01	8	0.5
1	11,662.00	1.40E+02	8	0.5
2	11,662.00	1.40E+02	0	0
3	11,662.00	1.40E+02	0	0
4	11,662.00	1.40E+02	0	0
5	11,662.00	1.40E+02	0	0
6	11,662.00	1.40E+02	0	0
7	11,662.00	1.40E+02	0	0
8	11,662.00	1.40E+02	0	0
9	11,662.00	1.40E+02	0	0
10	7,953.66	9.55E+01	0	0
11	3,725.99	4.47E+01	0	0
12	3,112.75	3.74E+01	0	0
13	2,710.46	3.25E+01	0	0
14	2,382.39	2.86E+01	0	0
15	2,111.00	2.53E+01	0	0
16	1,883.69	2.26E+01	0	0
17	1,691.18	2.03E+01	0	0
18	1,526.56	1.83E+01	0	0
19	1,384.58	1.66E+01	0	0
20	1,247.46	1.50E+01	0	0
21	0.00	0.00E+00	0	0
22	0.00	0.00E+00	0	0
23	0.00	0.00E+00	0	0
24	0.00	0.00E+00	0	0
25	0.00	0.00E+00	0	0
26	0.00	0.00E+00	0	0
27	0.00	0.00E+00	0	0
28	0.00	0.00E+00	0	0
29	0.00	0.00E+00	0	0
recovered	5.14E+07 stb	6.17E+05 mmscf		

Figure 7.9 Excel economics spreadsheet.

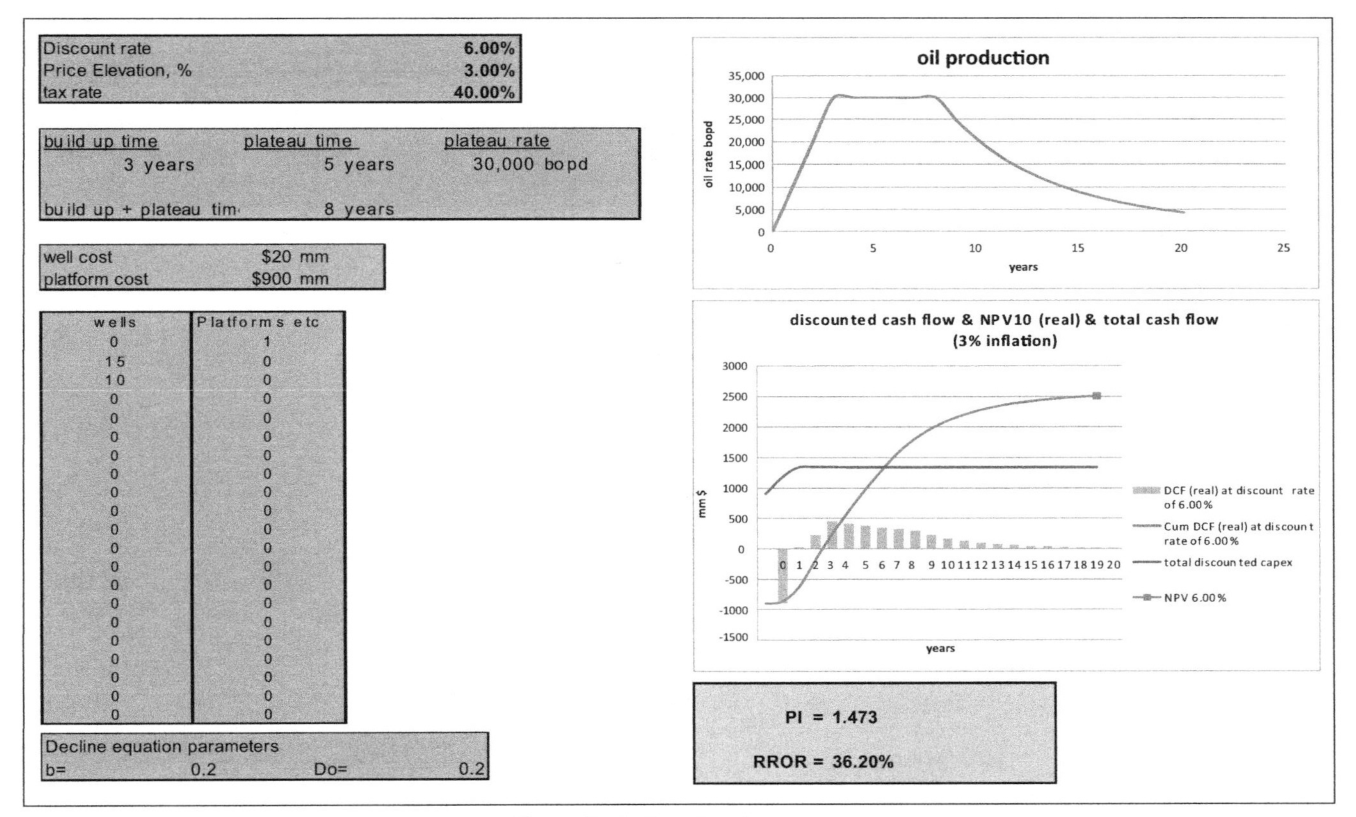

Figure 7.10 Case 1 — base case.

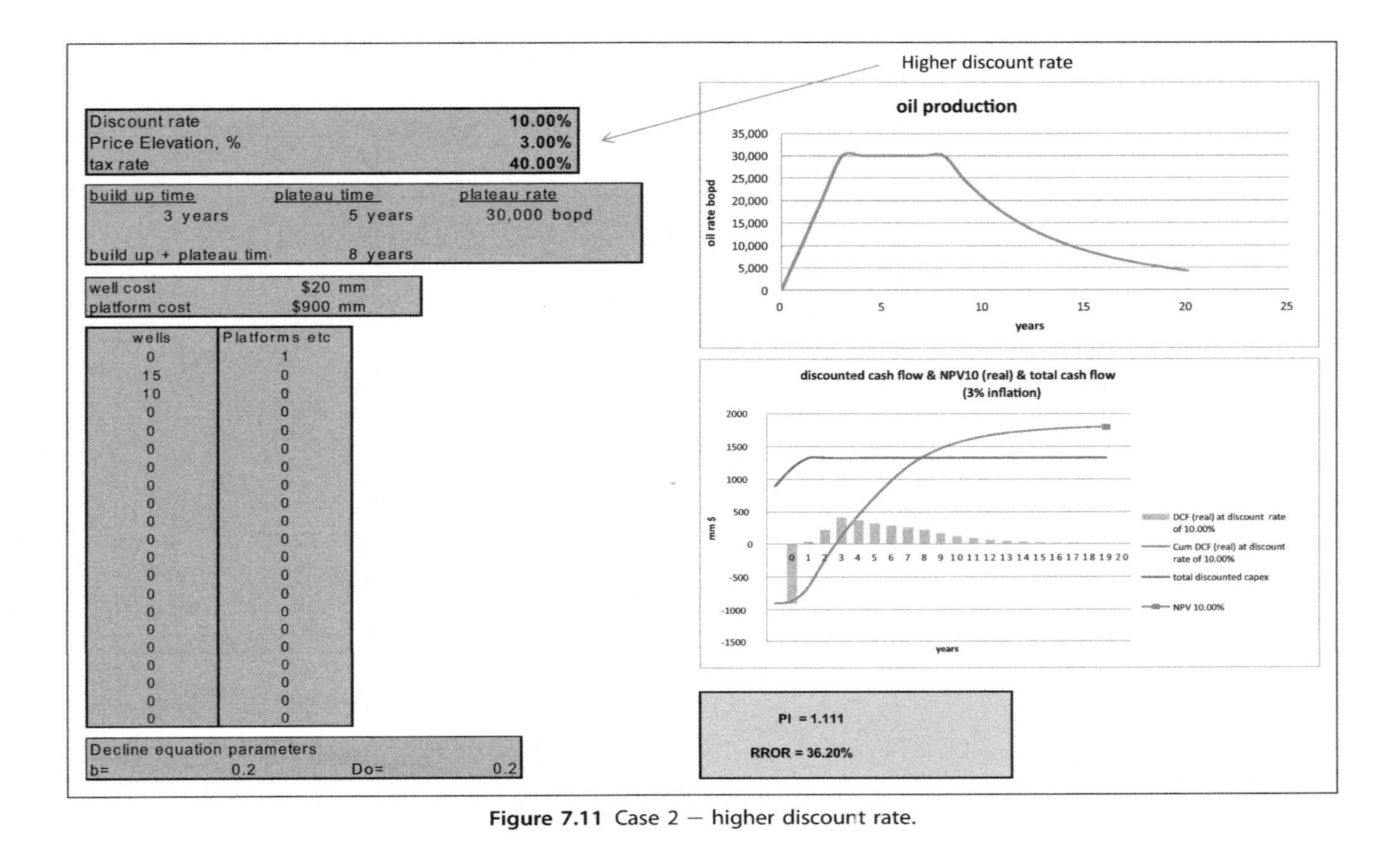

Figure 7.11 Case 2 — higher discount rate.

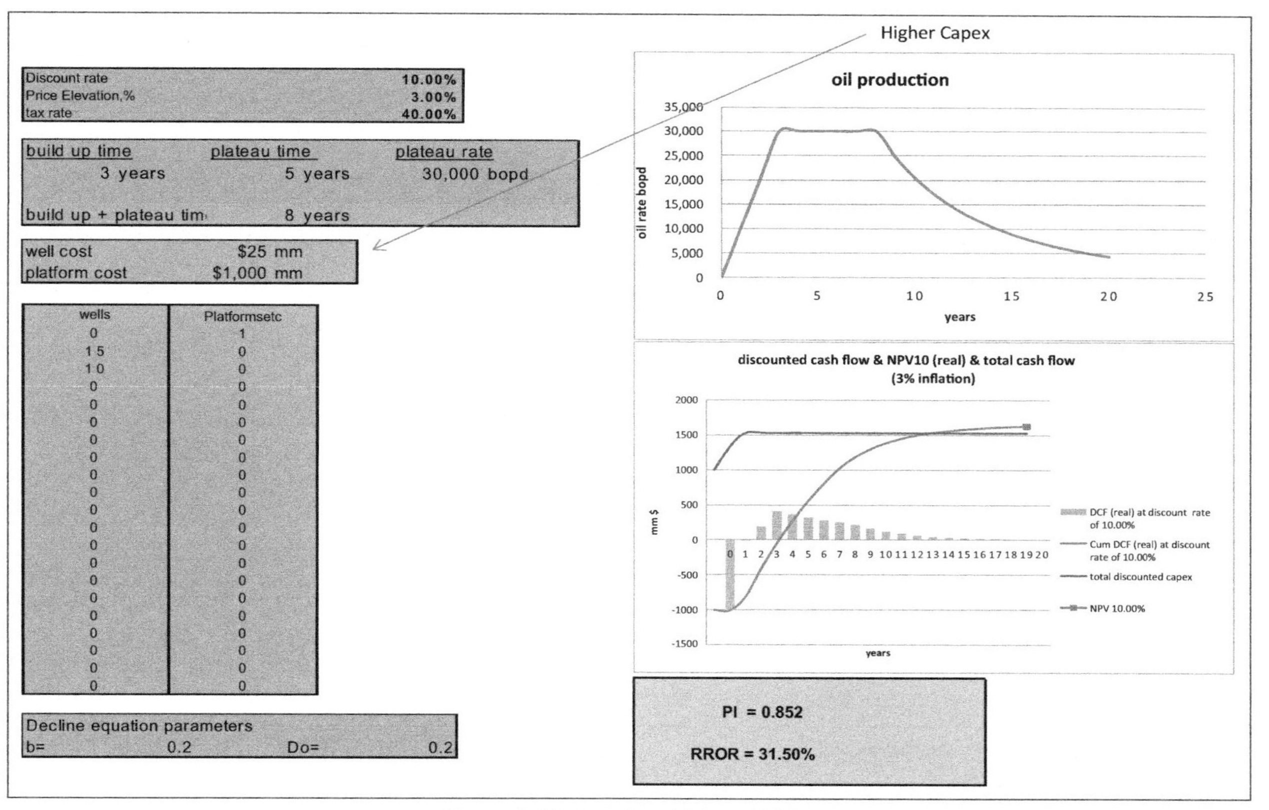

Figure 7.12 Case 3 — higher CAPEX.

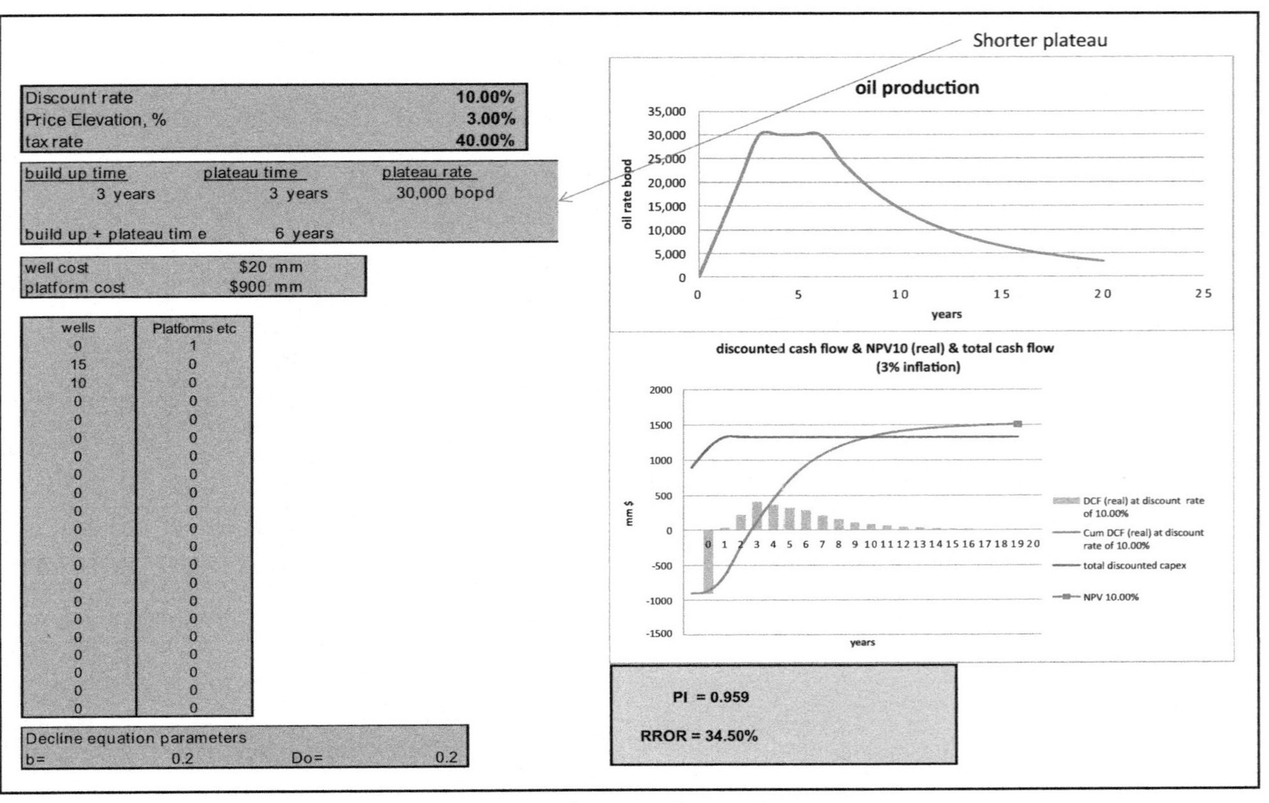

Figure 7.13 Case 4 — shorter plateau.

7.12 EFFECT OF VARIOUS PARAMETERS ON ECONOMIC INDICATORS

Fig. 7.14 shows the effect of discount rate, oil price, costs and reserves on NPV.

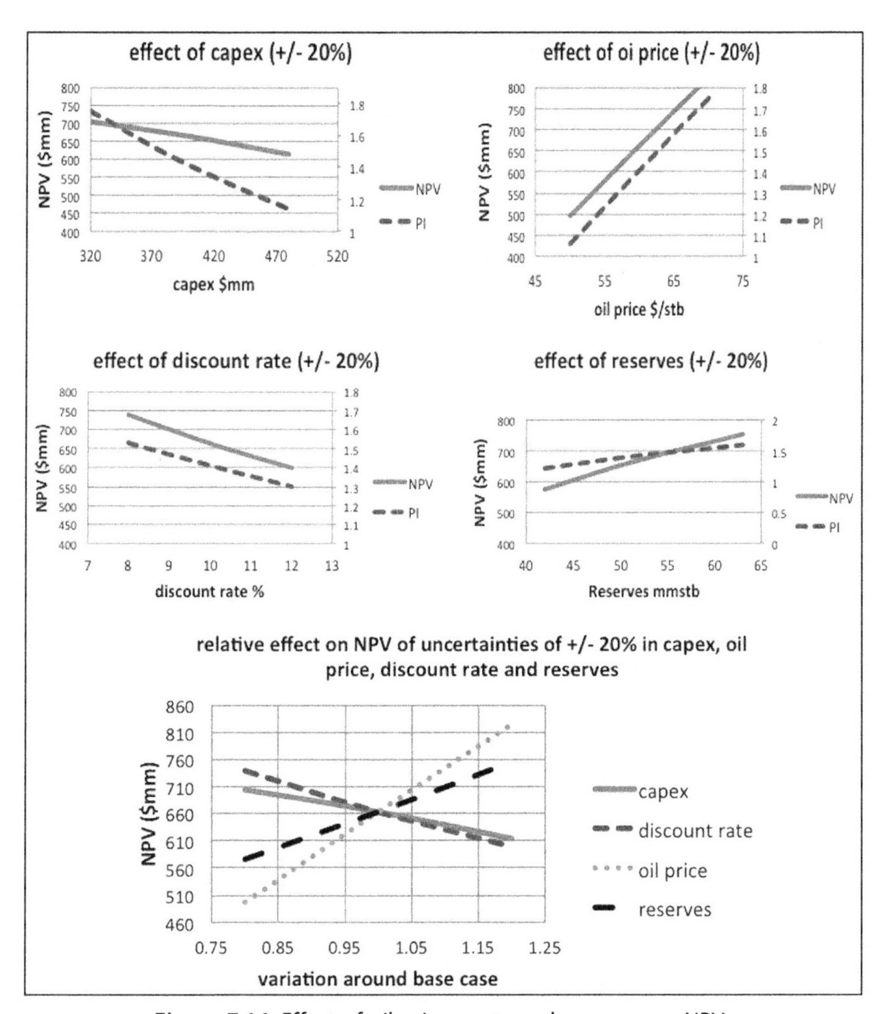

Figure 7.14 Effect of oil price, costs and reserves on NPV.

7.13 QUESTIONS AND EXERCISES

Q7.1. Explain the meaning of DCF and NPV.

Q7.2. The table below shows oil production and well drilling for a waterflood field.

Year	Wells (inj + prod)	Oil rate (bopd)
0	0	0
1	15	13,333
2	10	26,667
3	0	40,000
4	0	40,000
5	0	40,000
6	0	40,000
7	0	40,000
8	0	40,000
9	0	32,877
10	0	27,223
11	0	22,697
12	0	19,045
13	0	16,075
14	0	13,644
15	0	11,642
16	0	9,981
17	0	8,597
18	0	7,437
19	0	6,460
20	0	5,633
	Reserves	**168,379,042**

Facilities' costs are assumed to be spread evenly over three years. If wells cost $10 m each and total facilities' costs are $1.2 billion, use the software provided ("economic indicators") to calculate DCF, NPV and PI over 20 years of production assuming an oil price of $100/bbl, a discount rate of 10 per cent, taxation at 40 per cent and inflation at 3 per cent. Plot the results.

Use the software provided to determine the RROR.

Look at the effect of the following changes (keeping everything else constant).

1. Reducing discount rate to 6 per cent.
2. Decreasing the oil price to $80/bbl.
3. Increasing facilities' costs to $1.8 billion.

Q7.3. The table below shows oil production and well drilling for a gas field.

Year	Gas rate mmscf/d	Wells
0	0	0
1	100	6
2	200	5
3	300	0
4	300	0
5	300	0
6	300	0
7	300	0
8	300	0
9	300	0
10	186	0
11	121	0
12	81	0
13	56	0
14	40	0
15	29	0
16	21	0
17	16	0
18	12	0
19	9	0
20	7	0
	Reserves	**1.08 bcf**

Facilities' costs are assumed to be spread evenly over three years. If wells cost $10 m each and total facilities' costs are $1.6 billion, use the software provided ("economic indicators-gas") to calculate DCF, NPV and PI over 20 years of production, assuming a gas price of $10/mmscf, a discount rate of 10 per cent, taxation at 40 per cent and inflation at 3 per cent. Plot the results.

Use the software provided to determine the RROR.

Look at the effect of changing the following (keeping everything else constant).

1. Reducing discount rate to 6 per cent.

2. Decreasing the oil price to $8/scf.

3. Increasing facilities' costs to $2.2 billion.

Q7.4. The table below shows the oil production and drilling programme for downside (P90), base case (P50) and upside (P10) estimates for a field. If wells cost \$10 m each and total facilities' costs are \$1.6 billion (spread evenly over three years), calculate the NPV(10) (assuming 40 per cent taxation and 3 per cent inflation) for each case.

Year	Wells drilled	P50 oil rate bopd	P90 oil rate bopd	P10 oil rate bopd
0	6	0	0	0
1	8	10,000	10,000	10,000
2	8	20,000	20,000	20,000
3	0	30,000	30,000	30,000
4	0	30,000	30,000	30,000
5	0	30,000	30,000	30,000
6	0	30,000	30,000	30,000
7	0	30,000	24,658	30,000
8	0	30,000	20,417	30,000
9	0	24,658	17,023	30,000
10	0	20,417	14,283	24,658
11	0	17,023	12,056	20,417
12	0	14,283	10,233	17,023
13	0	12,056	8,731	14,283
14	0	10,233	7,486	12,056
15	0	8,731	6,448	10,233
16	0	7,486	5,578	8,731
17	0	6,448	4,845	7,486
18	0	5,578	4,225	6,448
19	0	4,845	3,697	5,578
20	0	4,225	3,247	4,845
	Reserves	**126 mmbbl**	**107 mmbbl**	**136 mmbbl**

Now calculate the EMV for the development, assuming probabilities of 25 per cent for P90, 50 per cent for P50 and 25 per cent for P10.

Q7.5. Explain the meaning of RROR.

Q7.6. Explain the difference between risked and unrisked economic indicators.

7.14 FURTHER READING

R.D. Seba, Economics of Worldwide Petroleum Production, Ogci & Petroskills Publications, 2008.

J. Masserson, Petroleum Economics, Editions Technip, 2000.
D. Johnston, International Exploration Economics, Risk and Contract Analysis, Penn Well, 2003.

7.15 SOFTWARE

economic indicators

CHAPTER 8

Field Appraisal and Development Planning

8.1 INTRODUCTION

This chapter discusses stages in the life of an asset—the exploration stage to discovery, the final investment decision, and on to production and abandonment—and the disciplines principally involved (Fig. 8.1).

The appraisal and development stage is absolutely critical in obtaining value from a discovered asset, and it is where reservoir engineers can have most influence on the key decisions. Early decisions have the greatest financial impact on a project. This is known as "front-end loading."

On discovery, decisions must be made on an initial appraisal programme—how many appraisal wells need to be drilled and where, and what testing is needed with these wells (log types, fluid sampling, laboratory studies, etc.)? Since we only have data from a single (discovery) well, this is probably not the time to build complex numerical models, but we still need to explore development options and potential asset value. There are four approaches here.

1. Use of analog field data.
2. Decline curve analysis.

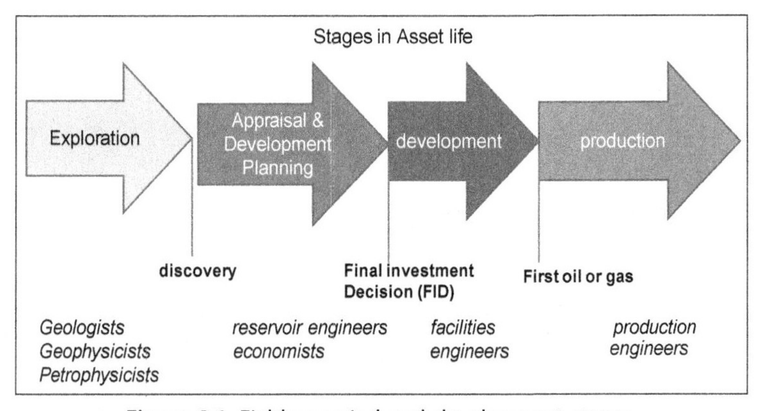

Figure 8.1 Field appraisal and development stages.

Fundamentals of Applied Reservoir Engineering
ISBN 978-0-08-101019-8
http://dx.doi.org/10.1016/B978-0-08-101019-8.00008-9
© 2016 Elsevier Ltd.
All rights reserved.

3. Analytical methods (eg, material balance or Buckley–Leverett-type analysis for water flooding).
4. Simple numerical models.

The first two of these are discussed in this chapter. Analytical methods and numerical modeling are covered in chapters "Estimation of Reserves and Drive Mechanisms" and "Fundamentals of Petroleum Economics".

Following initial appraisal, the need for further appraisal must be considered. A value of information (VOI) exercise based on a sensitivity analysis needs to be carried out. Alongside this, a detailed evaluation of potential developments based on current data, using either decline curve or a simple numerical model, is needed.

Early modeling of reservoirs, sensitivity analysis and VOI are examined in this chapter.

8.2 INITIAL EVALUATION OF POTENTIAL DEVELOPMENTS

Early evaluation of potential development schemes is necessary as an input to planning an appraisal programme which will be going on in parallel, and also future corporate financial commitments will need to be at least recognized even at an early stage (Fig. 8.2).

Therefore, given the very limited data available (from a single discovery well), a large number of simplifying assumptions will be necessary. We need to explore a range of options, particularly for oil fields (straight depletion, solution gas drive, water flooding) and gas condensate fields (depletion, gas

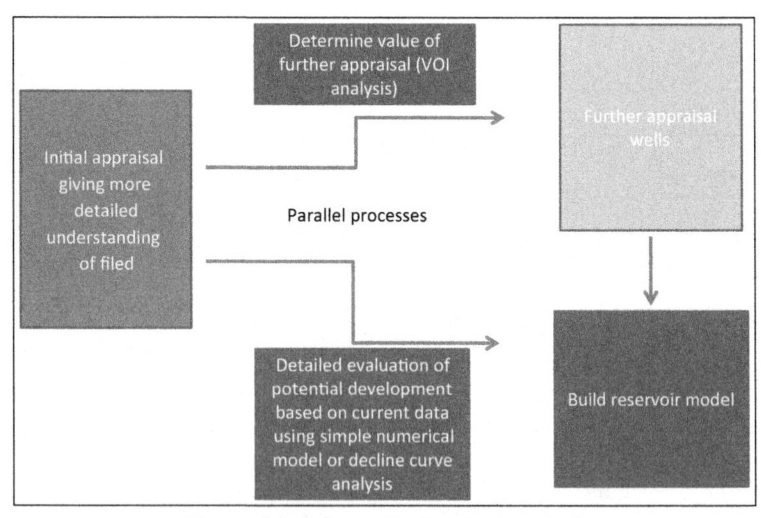

Figure 8.2 Schematic of early appraisal.

recycling). For each of these options, the nature of the production profile, well numbers, build-up time, and plateau rate need to be optimized for various economic indicators.

A methodology for obtaining a production profile for any particular development with limited early data is as follows.

1. Estimate reservoir oil or gas in place (V_o) by the normal method (porosity, area, reservoir thickness, net to gross, water saturation).
2. Note the initial flow rate (q_o), which will have to be an "unconstrained" rate—a constrained rate must be adjusted where necessary.
3. Make an initial assumption on recovery factor (R_f) based on reservoir type. Some typical values are shown in Table 8.1.
4. Take recoverable oil or gas from $V_R = V_o \cdot R_f$.
5. Make an initial assumption on well numbers (n_w).
6. We now need to obtain a single-well production profile. This can be done from analog data, using decline curve analysis or from a simple numerical model (either single-cell models as discussed in chapter "Analytical Methods for Prediction of Reservoir Performance" or a simple simulation model). Use of analog data and decline curves is discussed below. Single-well numerical modeling is discussed in chapters "Analytical Methods for Prediction of Reservoir Performance" and "Numerical Simulation Methods for Predicting Reservoir Performance".
7. Aggregate for total number of wells (or pairs of injector–producer wells with water flood) with well buildup, determining rate potential at any time and capping at plateau rate. Simple spreadsheets are available to do this ("aggregation oil and aggregation gas"). Single-well production rate, well timing, and capped plateau rate are input, and the resulting field rate is output. The spreadsheet balances the cumulative volume above the cap with that then available for additional production (see the example in Fig. 8.3).

   Examples of resulting capped profiles are shown in Fig. 8.4.

Table 8.1 Recovery factor ranges

Drive mechanism	Range (%)	Average (%)
Gas field—gas expansion	65–95	80
Oil field—oil expansion	2–10	6
Solution gas drive	25–35	30
Gas cap drive	20–40	30
Aquifer drive	20–40	30
Water flood	40–60	50

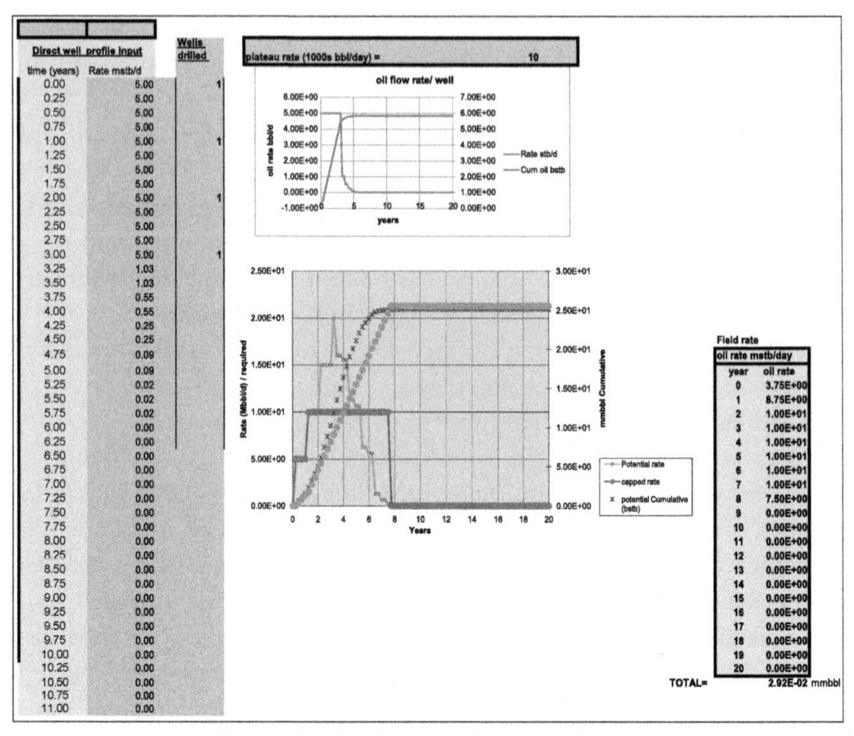

Figure 8.3 Excel aggregation spreadsheet.

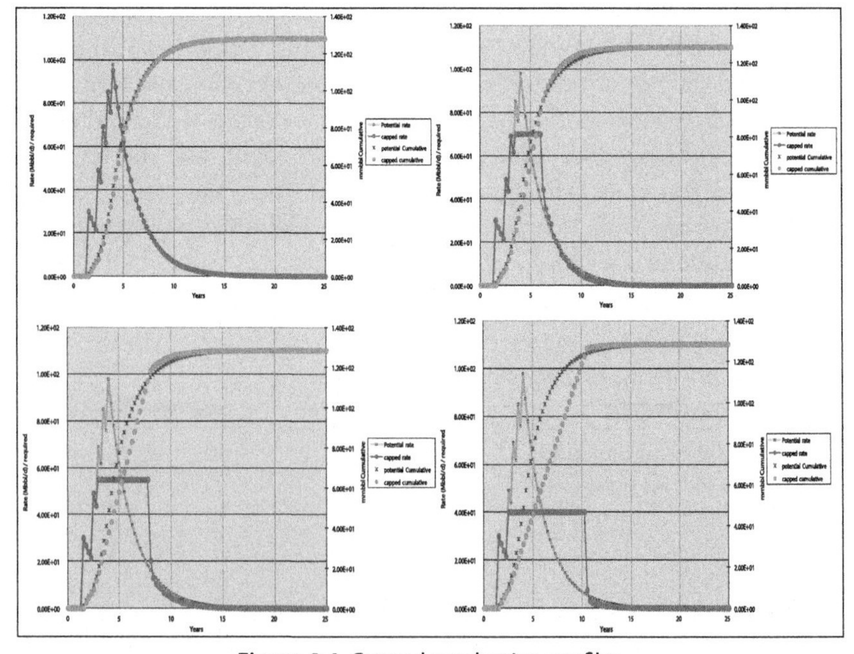

Figure 8.4 Capped production profiles.

8. Run economics to determine net present value (NPV), rate of return, etc.

   Determining factors will be as follows.

   a. Optimization of use of facilities. We will not want to build facilities (separation plant, etc.) with capacities that are only going to be used for a short period—some reasonable plateau rate must be determined.
   b. Political risk. Companies will extract hydrocarbons as rapidly as possible to get early return on investments in regions perceived as having high political risk.
   c. Contract arrangements. For gas in particular, sales contracts may be quite long, requiring a lower plateau rate—buyers may be more interested in long-term contracts at a lower rate.
   d. Reservoir characteristics. Reservoir uncertainty may suggest a lower (safer) plateau rate.
   e. Economic limit. Production will terminate when production costs equal the value of the product.

Development can now be optimized on the basis of all of these factors.

Calculation and significance of the various economic indicators are covered in an earlier chapter.

8.3 USE OF ANALOG DATA

Immediately after discovery, analog data from geologically similar fields, particularly if they are in the same region as the discovery, are very useful.

The analog field or fields need to have geological similarities in structure and also depositional similarities.

After discovery we can narrow down an analog field list.

Fields that have already been developed in the same way as that proposed for the discovered field are of course the most valuable. A recovery factor will be based on that achieved (or expected) in the analog field.

From available data a single-well profile must be generated.

8.4 EMPIRICAL DECLINE CURVE ANALYSIS

8.4.1 General

These are semiempirical equations with a sufficient number of variables to fit many types of typical well behavior.

Most commonly used is Arp's equation (Fig. 8.5):

$$q = q_o \frac{1}{(1 + b \cdot D_o t)^{1/b}} \quad [8.1]$$

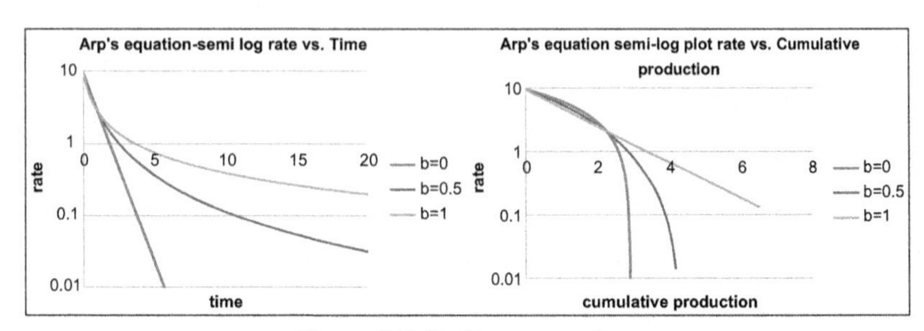

Figure 8.5 Decline curve plots.

where q_o = initial production rate, D_o = early decline parameter, and b = long-term decline parameter.

In the limit $b = 0$ this reduces to:

$$q = q_o e^{-D_o t}$$

Where $b = 1$ this becomes:

$$q = \frac{q_o}{1 + D_o t}$$

It must be made clear that any derivation of Arp's equation involves numerous assumptions, and the equation is therefore a gross simplification. It is, however, very useful in early modeling. *Initial well rate and early decline* can give valuable predictions for future production rates for fields.

Arp's equation must always be used with care, and considered as an empirical or at best a semiempirical equation.

8.4.2 Gas Wells

There is at least some case for treating Arp's equation as semiempirical.

For a single "free-flowing" well there will be three stages in decline behavior, assuming constant bottom hole pressure.

1. Initial decline flow.
2. Intermediate semisteady-state flow.
3. Boundary-dominated flow.

By changing the b factor in Arp's equation we can model the last two of these three stages approximately. For boundary-dominated flow, $b = 0$ (exponential decline) is appropriate. During the steady-state stage, $b = 1$ works well for gas wells. For initial decline, the parameter D_o is dominant so the value for b is much less important. The only case where "free-flow" modeling is normally needed is for shale gas wells, discussed in chapter "Unconventional Resources". Normally, for conventional fields, we will be modeling a collection of wells across a reservoir where well rates will be capped anyway until we reach a minimum bottom hole pressure when production goes into decline. At this point we will almost certainly be in boundary-dominated flow, and exponential decline ($b = 0$) should be used.

8.4.3 Oil Wells

Because of the number of different potential drive mechanisms and the strong dependence of reservoir factors involved with oil wells, Arp's equation should be considered as entirely empirical. Where we have a good amount of production data, Arp's equation can be fitted to these data for future prediction (provided the same drive mechanism is assumed for the future). In general, where no other guidance is available it is best to assume exponential ($b = 0$) decline. With water flooding we can assume a plateau rate close to the initial well potential rate q_o, where water injection maintains production before water breakthrough (see chapter: Estimation of Reserves and Drive Mechanisms). After this the safest assumption is again exponential decline.

8.4.4 Excel Spreadsheet for Arp's Decline Equation

Spreadsheets on Arps(oil) and Arps(gas) for producing decline curves are available.

8.5 USE OF SINGLE-WELL ANALYTICAL METHODS

These methods are covered in chapter "Analytical Methods for Prediction of Reservoir Performance", and include material balance, Buckley–Leverett and single-well time-dependent numerical models.

8.6 APPRAISAL PROGRAMME—SENSITIVITY ANALYSIS

Following the discovery well and early exploration of possible development options discussed above, a decision on initial appraisal wells needs to be made and also on what data need to be collected from these. This decision should be made on the basis of a sensitivity analysis determining the critical parameters that will affect production and cost profiles—this is best done by building a simple numerical reservoir model (discussed in Chapter 5) or

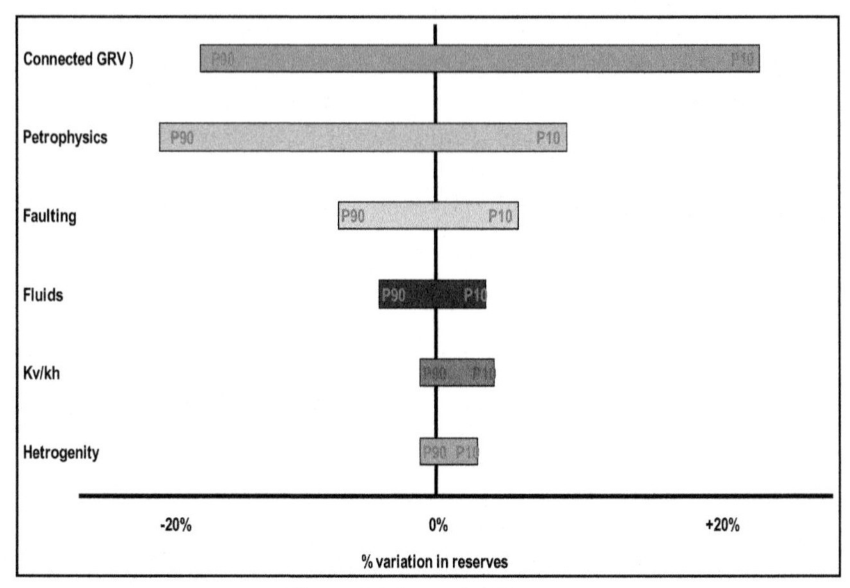

Figure 8.6 Tornado diagram.

the single-well/aggregation method discussed above. The results are then used to generate a "tornado diagram," an example of which is shown in Fig. 8.6.

Using this model and starting from the P50 (base case), P90 (downside), and P10 (upside) values of various parameters are input (all other parameters are kept constant) and the production profile and hence recoverable reserves are determined. As well as reserves, NPV can be used. Reserves and tornado diagrams are then constructed for each of these.

Some parameters such as gross rock volume (GRV in the example in Fig. 8.6) or petrophysical interpretation may have a very significant effect, while again in Fig 8.6 kv/kh (vertical/horizontal permeability) or reservoir heterogeneity may have a limited effect.

Appraisal wells and data-gathering decisions are then based on this analysis: wells and data giving more information on the most significant parameters should clearly be a priority. The actual value of gathering and analyzing these new data may be estimated using the VOI analysis described in the next section.

The "skew" in reserves resulting from upside or downside assumptions is also of importance in understanding reservoir risk.

8.7 VALUE OF INFORMATION

VOI is a cost/benefit exercise where we calculate estimated monetary values (EMVs) for potential situations where we invest, or not, in gathering further information to clarify the reservoir position. This is best explained with a simple example.

Consider a situation like that shown in Fig. 8.7. Here we have a field which has one discovery well and one appraisal well. It is not clear from seismic data if the right-hand side region of the field is connected across the fault (which may be sealing or only partly sealing), or if it has good permeability or not.

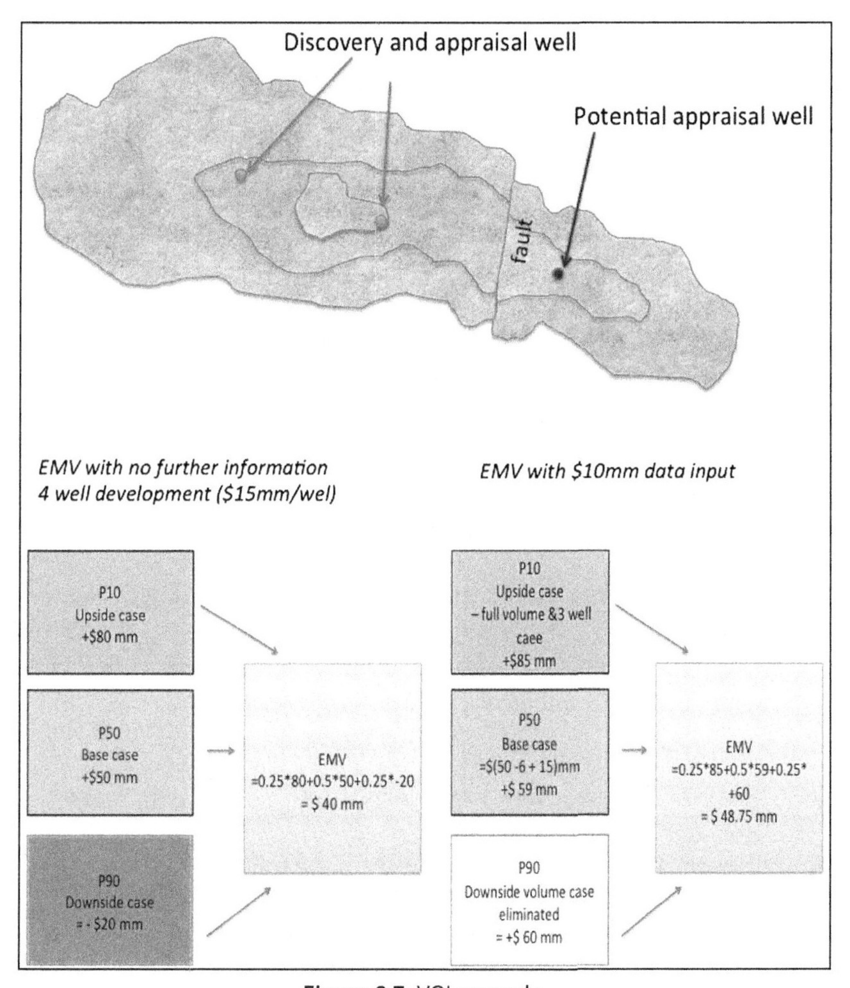

Figure 8.7 VOI example.

The question is, will we need four or just three development wells (saving $15 mm) for full field development, and is there sufficient volume in the reservoir to develop the field at all since the downside volume give a negative NPV.

What is the value of drilling a new appraisal well (cost = $10 mm)?

In the downside (P90) case the project has a negative NPV.

We estimate the EMV if the project goes forward with no further information on this parameter ($40 mm) and compare this with the EMV assuming "perfect information" from new data.

We assume four well developments costing $15 mm each and the cost of drilling an appraisal well of $10 mm.

If we discover that there is no effective reservoir in the undrilled region, the field will not be economic, so the "value" of this information is equal to the total cost of drilling up the field (4 × $15 mm = $60 mm). There is also a potential upside, where the new information allows the field to be developed with just three wells rather than four wells. We could treat this as a P10 case.

The final EMV in this "perfect information" case is $48.75 mm compared with $40.0 mm for the "no further information" case, so the VOI would be around + $8.75 mm (see Fig. 8.7).

It is important to be clear that new information *only has value if it will affect some monetary decision that can change as a result of the VOI exercise.*

8.8 QUESTIONS AND EXERCISES

Q8.1. List the stages of field development and the main disciplines involved in each.

Q8.2. The table below shows expected oil production rates (quarterly averaged) from a single production well in a water-flood field obtained by numerical sector modeling. We assume a line drive with an equal number of producer and injector wells.

years	Rate bbl/d	years	rate	years	rate	years	rate
0	0.00E+00						
0.25	1.00E+04	5.25	2.37E+02	10.25	2.70E+00	15.25	3.13E-02
0.5	1.00E+04	5.5	1.89E+02	10.5	2.16E+00	15.5	2.51E-02
0.75	1.00E+04	5.75	1.51E+02	10.75	1.73E+00	15.75	2.01E-02
1	1.00E+04	6	1.21E+02	11	1.38E+00	16	1.61E-02
1.25	1.00E+04	6.25	9.67E+01	11.25	1.10E+00	16.25	1.29E-02

years	Rate bbl/d	years	rate	years	rate	years	rate
1.5	1.00E+04	6.5	7.73E+01	11.5	8.83E-01	16.5	1.03E-02
1.75	1.00E+04	6.75	6.18E+01	11.75	7.07E-01	16.75	8.25E-03
2	4.40E+03	7	4.94E+01	12	5.66E-01	17	6.61E-03
2.25	3.51E+03	7.25	3.95E+01	12.25	4.53E-01	17.25	5.29E-03
2.5	2.81E+03	7.5	3.16E+01	12.5	3.62E-01	17.5	4.24E-03
2.75	2.24E+03	7.75	2.52E+01	12.75	2.90E-01	17.75	3.40E-03
3	1.79E+03	8	2.02E+01	13	2.32E-01	18	2.72E-03
3.25	1.43E+03	8.25	1.61E+01	13.25	1.86E-01	18.25	2.18E-03
3.5	1.14E+03	8.5	1.29E+01	13.5	1.49E-01	18.5	1.75E-03
3.75	9.12E+02	8.75	1.03E+01	13.75	1.19E-01	18.75	1.40E-03
4	7.28E+02	9	8.24E+00	14	9.52E-02	19	1.12E-03
4.25	5.82E+02	9.25	6.59E+00	14.25	7.62E-02	19.25	8.98E-04
4.5	4.65E+02	9.5	5.27E+00	14.5	6.10E-02	19.5	7.19E-04
4.75	3.71E+02	9.75	4.22E+00	14.75	4.88E-02	19.75	5.76E-04
5	2.97E+02	10	3.37E+00	15	0.03907925		

The reservoir has a hydrocarbon pore volume of 200 mmbbl. Use the aggregation and oil field economics software provided to optimize a development (in terms of NPV(10)) by varying well timing and plateau rate. Assume a 12-producing well (+12 injectors) development, wells cost of $20 mm each and the cost of facilities needed for various maximum oil rates as shown below.

Maximum rate (bbl/day)	Cost $mm
10,000	800
20,000	900
30,000	1000
40,000	1300
50,000	1600
60,000	1900
70,000	2200
80,000	2500
90,000	2800
100,000	3100

Inflation is assumed to be at 3%, taxation at 40%, and oil price $90/bbl. Range drilling schedule as 3 wells/year, 6 wells/year, and 12 wells/year.

Q8.3. For the optimum case from Q8.2, examine the effect of "worse than expected" average well performance occurring. The final reserves are approximately the same as the above case, but the wells come off plateau sooner with a slower decline thereafter. This actual well performance is shown below.

Years	Rate bbl/d						
0.00	0.00E + 00						
0.25	1.00E + 04	5.25	1.50E + 02	10.25	1.71E + 00	15.00	2.49E − 02
0.50	1.00E + 04	5.50	1.20E + 02	10.50	1.37E + 00	15.25	1.99E − 02
0.75	1.00E + 04	5.75	9.59E + 01	10.75	1.10E + 00	15.50	1.59E − 02
1.00	1.00E + 04	6.00	7.66E + 01	11.00	8.77E − 01	15.75	1.28E − 02
1.25	5.46E + 03	6.25	6.13E + 01	11.25	7.01E − 01	16.00	1.02E − 02
1.50	4.36E + 03	6.50	4.90E + 01	11.50	5.61E − 01	16.25	8.19E − 03
1.75	3.48E + 03	6.75	3.91E + 01	11.75	4.49E − 01	16.50	6.56E − 03
2.00	2.78E + 03	7.00	3.13E + 01	12.00	3.59E − 01	16.75	5.25E − 03
2.25	2.22E + 03	7.25	2.50E + 01	12.25	2.88E − 01	17.00	4.21E − 03
2.50	1.77E + 03	7.50	2.00E + 01	12.50	2.30E − 01	17.25	3.37E − 03
2.75	1.42E + 03	7.75	1.60E + 01	12.75	1.84E − 01	17.50	2.70E − 03
3.00	1.13E + 03	8.00	1.28E + 01	13.00	1.47E − 01	17.75	2.16E − 03
3.25	9.04E + 02	8.25	1.02E + 01	13.25	1.18E − 01	18.00	1.73E − 03
3.50	7.22E + 02	8.50	8.18E + 00	13.50	9.45E − 02	18.25	1.39E − 03
3.75	5.77E + 02	8.75	6.54E + 00	13.75	7.56E − 02	18.50	1.11E − 03
4.00	4.61E + 02	9.00	5.23E + 00	14.00	6.05E − 02	18.75	8.92E − 04
4.25	3.68E + 02	9.25	4.18E + 00	14.25	4.85E − 02	19.00	7.15E − 04
4.50	2.94E + 02	9.50	3.34E + 00	14.50	3.88E − 02	19.25	5.73E − 04
4.75	2.35E + 02	9.75	2.68E + 00	14.75	3.11E − 02	19.50	4.59E − 04
5.00	1.88E + 02	10.00	2.14E + 00			19.75	3.68E − 04

Calculate the NPV(10) for this case and compare with the above.

Q8.4. Use the software ("Arp's equation-oil") to obtain plots assuming an initial rate of 10 mmbbl/day, 50% decline in the first year of production and b factors of 0.0, 0.5, and 1.0. Compare decline of production, rate and log(rate) versus time and log of rate versus cumulative production.

Q8.5. The table below shows the P90−P10 range of uncertainty on a number of reservoir parameters. Use the software provided ("Tornado diagram") to generate a tornado diagram based on these data. Discuss any skew in the results.

	Base case reserves (mmbbl)		99.6
	P90	**P10**	
Seismic GRV	52	120	
Petrophysics	77	110	
Faults	89	114	
Kv/kh	92	104	
Heterogeneity	88	100	
PVT	92	102	

Q8.6. We have discovered a small gas field where the major uncertainty is the gas/water contact depth over a significant portion of the reservoir. The best estimate P50 NPV(10) is $180 mm with an upside P10 of $200 mm, but there is a very significant downside P90 of $90 mm due to the risk of water ingress and conning if the gas/water contact is higher than expected. An appraisal well at a cost of $10 mm would clarify the issue. If the gas/water contact was at the higher level, the problem could be overcome by drilling highly horizontal wells. The additional cost of this would be $15 mm. Use the provided software ("VOI") to determine the value of drilling the appraisal well. We can assume that horizontal wells will completely remove the P90 risk.

Q8.7. A small offshore oil field has just been discovered. As the reservoir engineer, you have been asked to examine a potential water-flood development option based on the very limited data available. These are summarized below.

Geological/petrophysical data:

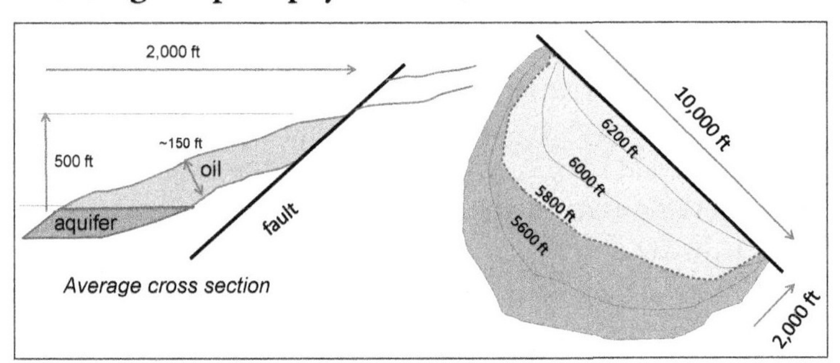

Average porosity = 15%, connate water saturation = 20%, net to gross = 0.80.

The discovery well flowed at 4500 stb/day at a drawdown of 1000 psi.

Laboratory data:

Absolute permeability = 150 mD.

Relative permeabilities are shown below.

S_w	k_{rw}	K_{ro}
0.20	0.00	0.90
0.25	0.001	0.73
0.30	0.01	0.58
0.35	0.02	0.44

Continued

S_w	k_{rw}	K_{ro}
0.40	0.04	0.32
0.45	0.06	0.23
0.50	0.10	0.14
0.55	0.14	0.08
0.60	0.20	0.04
0.65	0.27	0.01
0.70	0.35	0.00
0.75	0.44	0.00
0.80	0.55	0.00

Oil viscosity = 2.0 cP, water viscosity = 0.5 cP, Bo = 1.5, specific gravity of oil = 0.75, specific gravity of water = 1.05.

Economic assumptions: well costs = $10 mm/well, facilities cost = $100 mm per 2000 stb/day capacity, oil price = $100/bbl, discount rate = 10%, inflation rate = 3%, and taxation rate = 40%.

You firstly need to consider the location of wells (production and injection) to water flood the reservoir to obtain an estimated single-well production profile. It is suggested that you use the spreadsheet available ("waterflood") and input some of the above geological and laboratory data. Assume a four injector +4 producer well development. Water injection rate should be assumed to be 6000 stb/day.

With the single-well production profile, the total field profile will need to be optimized (timing and field plateau rate), Again, it is suggested that you use the available spreadsheets ("aggregation-oil" and "economics indicator").

Put together your optimized development plan along with your reasoning for all stages and the spreadsheet input and output. Show the well distribution you propose, the single-well production profile with recovery factor at breakthrough, and the final recovery factor. You need to show why this has the best economics. Show how other development possibilities (well timing and field plateau rates) give suboptimal economics.

8.9 FURTHER READING

R. Brafvoid, E. Bickel, H. Lohne, Value of Information in Oil and Gas Industry, 110,378-PA SPE Journal Paper (2009).

P. Cockcroft, K. Moore, Development Planning: A Systematic Approach, SPE 28,782 (1994).

8.10 SOFTWARE

Aggregation-oil
Aggregation-gas
Arp's equation (oil)

CHAPTER 9

Unconventional Resources

9.1 INTRODUCTION

Unconventional gas and oil resources are becoming an increasingly important part of hydrocarbon resources. In this chapter we look at the following resources.

1. Shale gas and oil.
2. Coalbed methane (CBM—known outside the United States as coal seam gas).
3. Heavy oil.

Shale gas/oil in particular is becoming increasingly important, as illustrated in Fig. 9.1.

9.2 DIFFERENCES BETWEEN CONVENTIONAL AND UNCONVENTIONAL RESOURCES

The major differences between conventional and unconventional resources common to shale gas and oil and coal seam gas are summarized in Table 9.1.

Unconventional resources normally exist in accumulations that are pervasive over large areas and are not significantly affected by hydrodynamic influences, thus reservoir limits are more difficult to establish. There are often extreme variations in coal or shale properties over these areas, which means that high sampling density and pilot schemes are more likely to be required than in conventional fields in an effort to reduce uncertainty when undertaking major developments.

9.3 SHALE GAS AND OIL

9.3.1 Global Distribution

Known global shale gas basins are shown in Fig. 9.1. US shale gas and oil is currently the most widely developed. Shale oil reserves are present in the United States mainly in the Bakken and Eagle Ford formations. Development of shale gas and oil has recently encountered many environmental concerns, which have slowed development both in the United States and elsewhere.

Fundamentals of Applied Reservoir Engineering
ISBN 978-0-08-101019-8
http://dx.doi.org/10.1016/B978-0-08-101019-8.00009-0
© 2016 Elsevier Ltd.
All rights reserved.

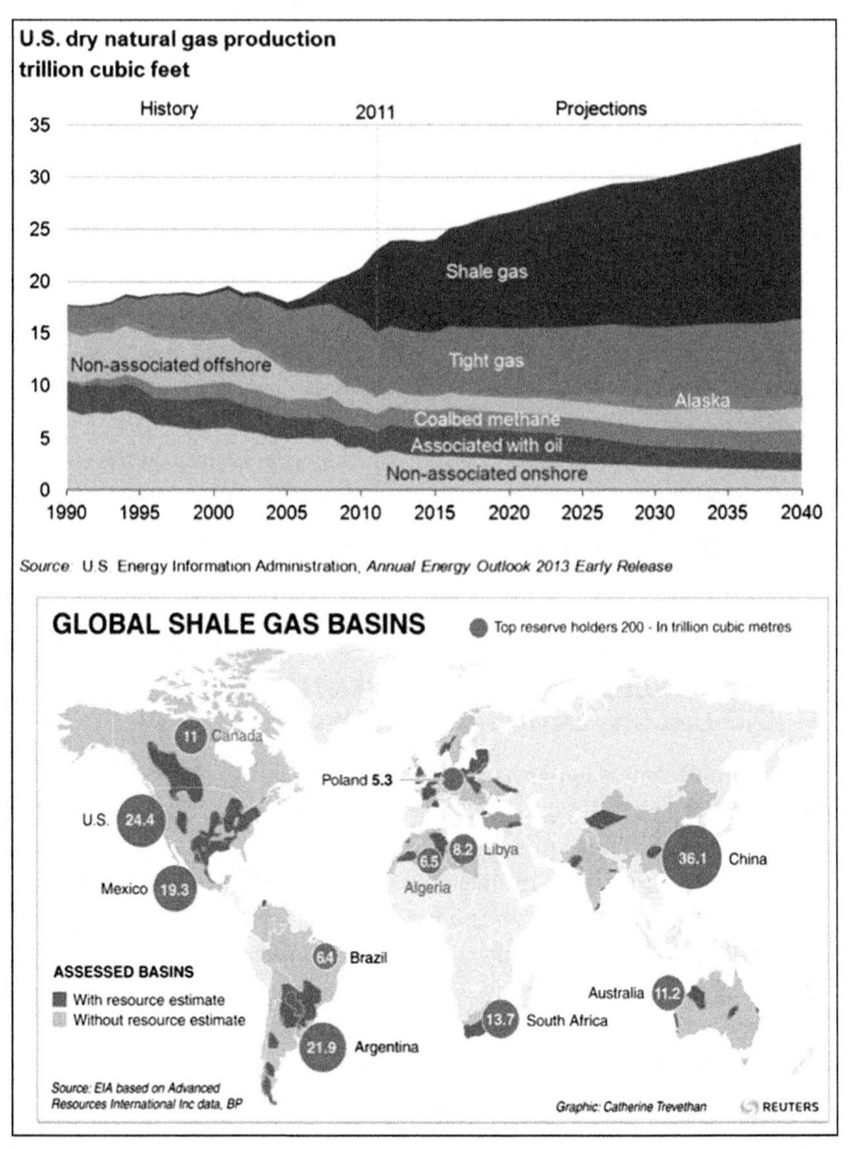

Figure 9.1 Gas production by source type and worldwide shale gas resources. *(The United States Energy Intelligence Agency.)*

9.3.2 Nature of Shale

Shale is a fine-grained sedimentary rock composed of mud (a mix of flakes of clay and fragments of quartz and calcite). It has very low permeability (in the nano (10^{-9}) Darcy range—Fig. 9.2), but it may have

Table 9.1 Summary of main differences between conventional and unconventional resources

	Conventional	Unconventional
1	Discrete accumulation	Continuous accumulation
2	Free gas or oil saturation	Free gas or oil plus adsorbed hydrocarbon
3	Low to moderate variation in permeability	High to extreme permeability variation
4	Increasing water production	Decreasing water production
5	Typically modeled as continuous reservoir	Typically best modeled as aggregation of single wells (due to permeability variation)
6	Normally relatively small number of wells (10–100)	Normally large number of wells (>100)
7	Pilot development not typical	Pilot development normal

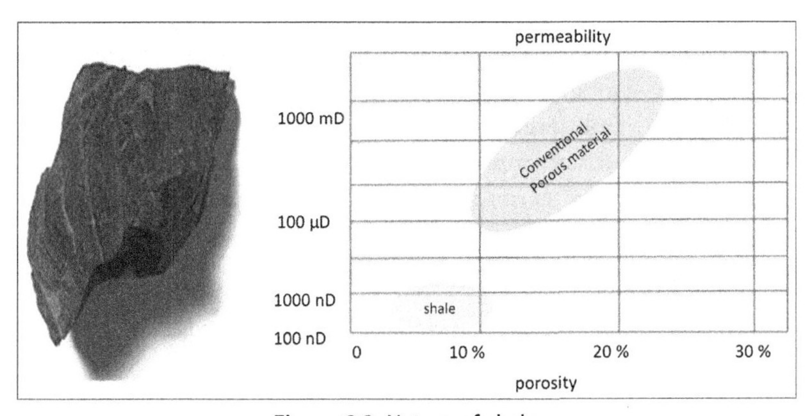

Figure 9.2 Nature of shale.

significant organic content (known at the total organic content—TOC), measured by gamma ray output on logs (2% is normally considered to be a minimum for a viable shale resource).

In shale gas and oil, hydrocarbons are generated "in situ" so that the shale serves as a source, a reservoir, and a seal, unlike the situation in conventional reservoirs.

High-quality shale will need to have both good porosity and significant TOC.

Gas in place (GIP) is calculated on a per acre basis for TOC, porosity, and thickness of the shale layer. Some gas is in an adsorbed state, and can be freed at lower pressure and added to the total GIP.

Because of its very low permeability the shale must be fractured so that gas (or oil) can be produced.

9.3.3 Fracking

Due to the very low permeability, hydraulic fracturing is normally required to make wells productive. Hydraulic fracturing, commonly known as fracking, is a technique in which water is mixed with sand and chemicals, and the mixture is injected at high pressure into a wellbore to create fractures along which fluids such as gas and water may migrate to the well (Fig. 9.3). The small grains of propant (sand or aluminum oxide) injected hold these fractures open once the rock achieves equilibrium.

Typically shale gas wells are horizontal (4000–5000 ft), with 8–12 fracs.

There are effectively two regions opened up by fracking.

A **hydraulically fractured volume** (HFV) is directly accessed by the injected fluid and sand that will be "propped" (held open). These fractures are normally vertical.

A **stimulated fractured volume** (SFV) is unpropped and induced beyond the hydraulically fractured region. It is typically horizontal, and originates from preexisting microfractures.

A total **accessed volume** (HFV + SFV) is opened up in fracking, and this, along with the permeability (combined effective fractures + shale matrix able to feed into the fractures), will determine the ultimate recovery factor (RF).

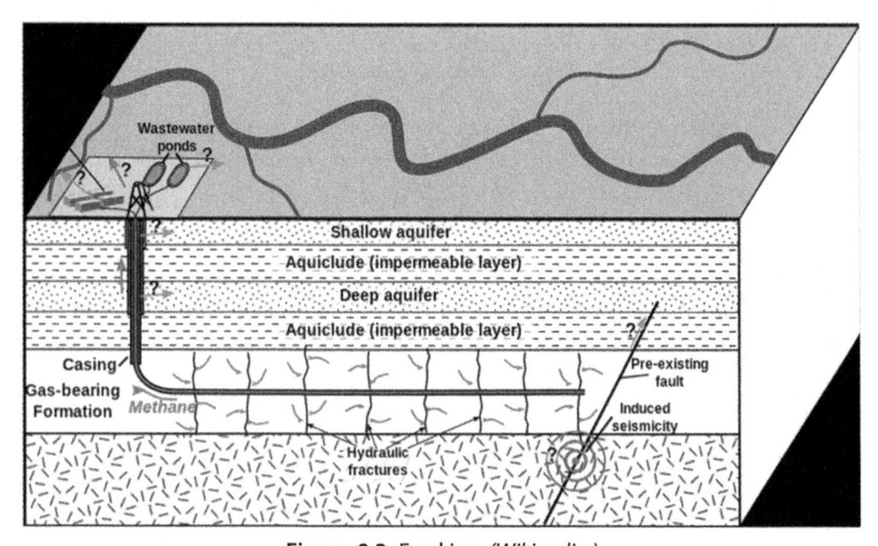

Figure 9.3 Fracking *(Wikipedia.)*

How we reach the recovery factor will also be a function of the **fracture geometry** and the initial reservoir pressure (dependent mainly on depth).

Successful fracking will depend on two factors.

1. The nature of microfractures and other fractures in the shale prior to fracking into which the hydraulic fractures spread.
2. The geomechnical properties of the shale itself—brittle, etc. (high or low Young's modulus). Fracking will normally be more successful with brittle shales.

Following fracking, wells can have initial production rates in the range of 1−12 mmscf/day. Success will depend on the shale matrix/fracture interface opened up and the fracture permeability established.

9.3.4 Use of Microseismic to Monitor Fracture Stimulation

Microseismic is used to detect and locate microearthquakes induced by hydraulic fracturing, and thus the areal and vertical extent of fracture induction. We can then map the geometry of these fractures. Sensitive seismometers are placed in adjacent wells. At each fracturing, resulting seismic events are recorded so that the extent of the secondary fracturing can be established. A schematic of this is shown in Fig. 9.4.

9.3.5 Shale Gas Reserves

For shale, GIP is normally quite well defined from early log data.

Reserves are simply the standard oil in place or GIP times the RF, so for a shale oil reservoir:

$$R = 7758\ A\ h_v \varphi (1 - S_w)/B_{oi} \cdot \mathrm{RF}\ \text{stb} \quad [9.1]$$

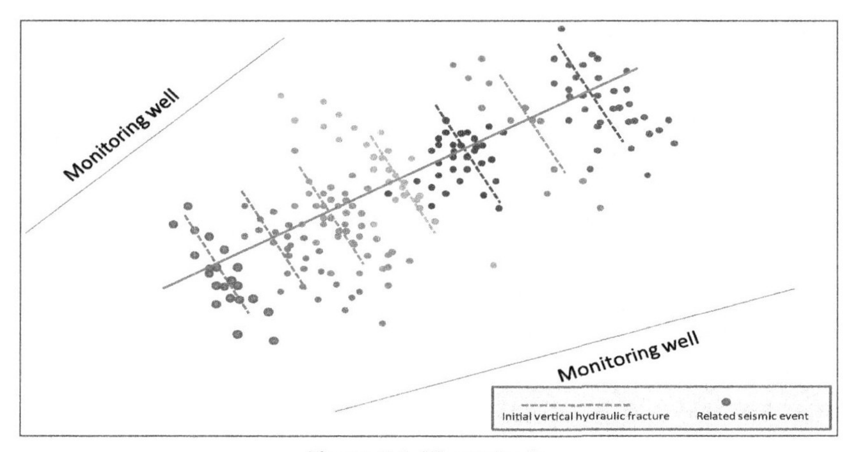

Figure 9.4 Microseismic.

For a shale gas reservoir:

$$R = 7758\ A\ h_v \varphi (1 - S_w)/B_{gi} \cdot \mathrm{RF}\ \mathrm{scf} \quad [9.2]$$

The key questions therefore are the technical RF and the economics of a given development project (ie, the cost of drilling and success of fracking). The technical RF will depend on the accessed volume (discussed above), which is very difficult to estimate until we have a significant amount of production data from a sample of wells.

RFs for shale gas are typically between 10% and 30% (over a 30-year period), known as the estimated ultimate recovery, so for example 1300 tcf GIP could give an estimated ultimate recovery of 130–390 tcf of recoverable gas.

Equally important, however, are the economics. Shale gas/oil wells can cost between $4 million and $9 million to drill and fracture, so the gas price is critical.

Oil shale tends to be more economic than gas shale, but in either case a certain rate of return has to be expected before any project will go forward.

Modeling shows that typically 10–30% of GIP is accessed by fracking, depending on fracture density, fracture surface area/matrix volume, and how much gas in the shale matrix can migrate into the fractures in given period of time. So accessed volume will depend on natural fractures in shale, stresses in the system, and hydraulic fracturing methods/conditions.

9.3.6 Estimation of Production Profiles

Shale gas wells can initially produce at quite high rates (up to 12 mmscf/d), but production declines rapidly (by as much as 80% within the first year). The Arp's (or hyperbolic) equation discussed in chapter "Field Appraisal and Development Planning" is normally used to estimate expected individual well performance. This is then aggregated for total shale gas field predictions. However, Arp's equation is in fact not really suitable for predicting shale gas decline and will often overpredict estimated ultimate recovery. The power law equation of Ilk and others, provided a suitable set of parameters is used, is more suitable for production and reserves predictions. Simple numerical models (such as dual-porosity numerical simulation models) are also useful tools with shale gas and oil, and remove the need to use semiempirical equations such as Arp's or power law that can be unreliable.

9.4 COALBED METHANE

9.4.1 Global Distribution

Current development of CBM is mainly in the United States, Canada, and Australia.

9.4.2 Nature of CBM

CBM is methane **adsorbed** (as a single molecular layer) on to the solid matrix of coal. The open fractures in the coal (called cleats and normally roughly orthogonal) can also contain free gas, but are normally initially saturated with water. This is quite distinct from conventional gas reservoirs, where the gas is initially in the pore space of the gas region.

Unlike much natural gas from conventional reservoirs, CBM contains very few heavier hydrocarbons. It often contains up to a few percent of carbon dioxide.

When drilled, water is produced and the pressure in the cleats drops. As pressure decreases the amount of methane that can be held on the coal matrix surfaces decreases and gas is evolved, thus water production decreases and gas production increases with time (Fig. 9.5). The amount desorbed will depend on the nature of the Langmuir isotherm, as shown in Fig. 9.5.

9.4.3 Estimation of Gas in Place

GIP in CBM is estimated from the area, the net coal-seam thickness, the in situ density of the coal, and the in situ gas content, with the ash and water content discounted. So:

$$\mathrm{GIP} = A \cdot h_{\mathrm{v}} \cdot \rho_{\mathrm{c}} \cdot G_{\mathrm{c}}(1 - A_{\mathrm{c}} - W_{\mathrm{c}}) \quad [9.3]$$

or in field units:

$$\mathrm{GIP} = 4.36 \times 10^{-5} \cdot A \cdot h_{\mathrm{v}} \cdot \rho_{\mathrm{c}} \cdot G_{\mathrm{c}}(1 - A_{\mathrm{c}} - W_{\mathrm{c}}) \quad [9.4]$$

where GIP = gas in place (bcf) in reservoir conditions; A = area (acres); h_{v} = net coal thickness (ft); ρ_{c} = in situ coal density (gm/cm^3); G_{c} = in situ gas content (sm^3/ton); A_{c} = ash content of coal (fraction); and W_{c} = moisture content (fraction).

9.4.4 Recovery Factors

Maximum recovery will depend on the nature of the Langmuir isotherm.

The maximum RF is given by:

$$\mathrm{RF} = (\text{Initial gas content} - \text{Final gas content})/\text{Initial gas content}$$

Typical RFs are in the range of 30–50%.

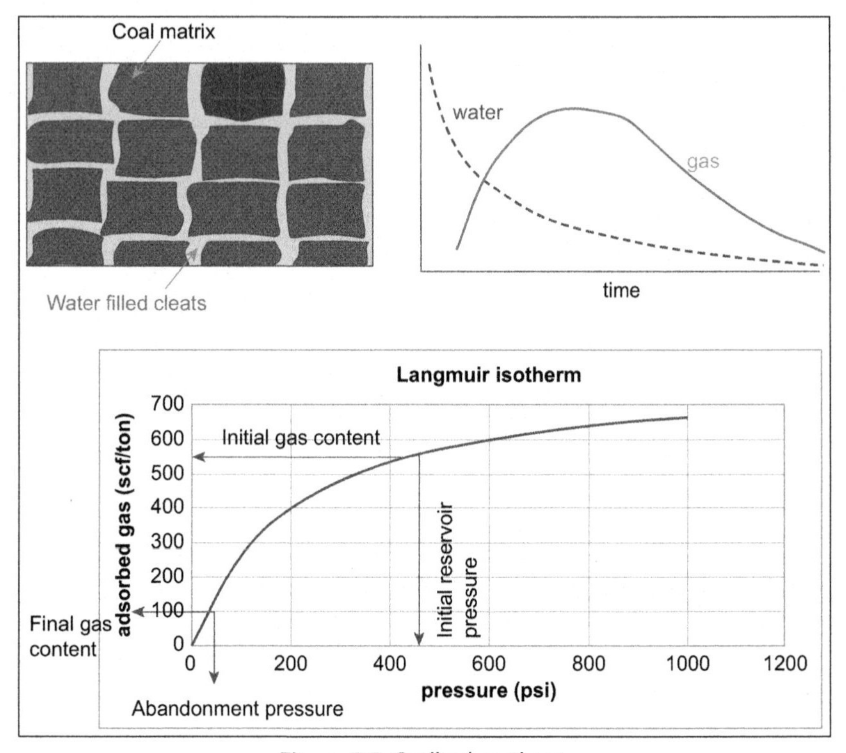

Figure 9.5 Coalbed methane.

9.4.5 Estimation of Production Potential and Reserves

From the above:

$$R = 4.36 \times 10^{-5} \cdot A \cdot h_v \cdot \rho_c \cdot G_c (1 - A_c - W_c) ^* \mathrm{RF} \qquad [9.5]$$

The above RF determined from the Langmuir isotherm is actually an ideal recovery—the most that can be achieved for the isotherm assumed. Complete and effective drainage of water is assumed, and this can often be a problem with CBM wells. An active aquifer can partially or completely prevent the pressure drop needed to release the gas. For this reason pilot schemes are normally needed to evaluate the risk.

There are also problems due to the extreme well-to-well variability in reservoir properties in many cases. Assumption of homogeneous properties over significant regions can therefore be misleading. Decline curves (normally the Arp's equation) for a single "average" well, adapted to allow for flow rate buildup with cleat water production decline, are often used in modeling CBM. This can be particularly useful when sufficient averaged well history is available. Production rates build up slowly as water is expelled. Typical maximum rates are normally below 1.0 mmscf/d, but, unlike the position with shale gas, production decline is relatively slow.

CBM reservoirs are also modeled using dual-porosity type numerical models with the addition of Langmuir isotherms to reproduce desorption from the coal matrix.

9.4.6 Water Disposal

The early production of water can present a problem in developing large-scale coal-seam gas resources. Options for disposal are collection "ponds" where evaporation gradually reduces water volumes; discharge into rivers; and reverse osmosis. The first two options are not environmentally favorable, as the water often contains significant amounts of salt and/or sodium bicarbonate, but purification as in reverse osmosis is expensive.

9.5 HEAVY OIL

9.5.1 General

Heavy oils are normally classified as oils with less than 20°API (American Petroleum Institute density definition) and viscosities greater than 200 cP at reservoir conditions. As conventional oil resources are gradually depleted, extraction of heavy oils will become increasingly important. Worldwide, heavy oil resources are probably up to twice those of conventional oil.

Traditional methods of extraction are steam injection and combustion. A new and as yet largely untested method is "cold heavy oil production with sand."

These methods are briefly discussed below. The traditional methods mainly make use of the strong dependence of heavy oil viscosity on temperature. A typical plot of this dependence is shown in Fig. 9.6(a). Interfacial tension is also reduced.

9.5.2 Continuous Steam Injection

There are two methods, continuous steam injection and cyclic steam injection.

Continuous steam flooding is comparable to water injection (discussed above), so steam injection provides drive as well as reducing oil viscosity. The schematic in Fig. 9.6 shows this process.

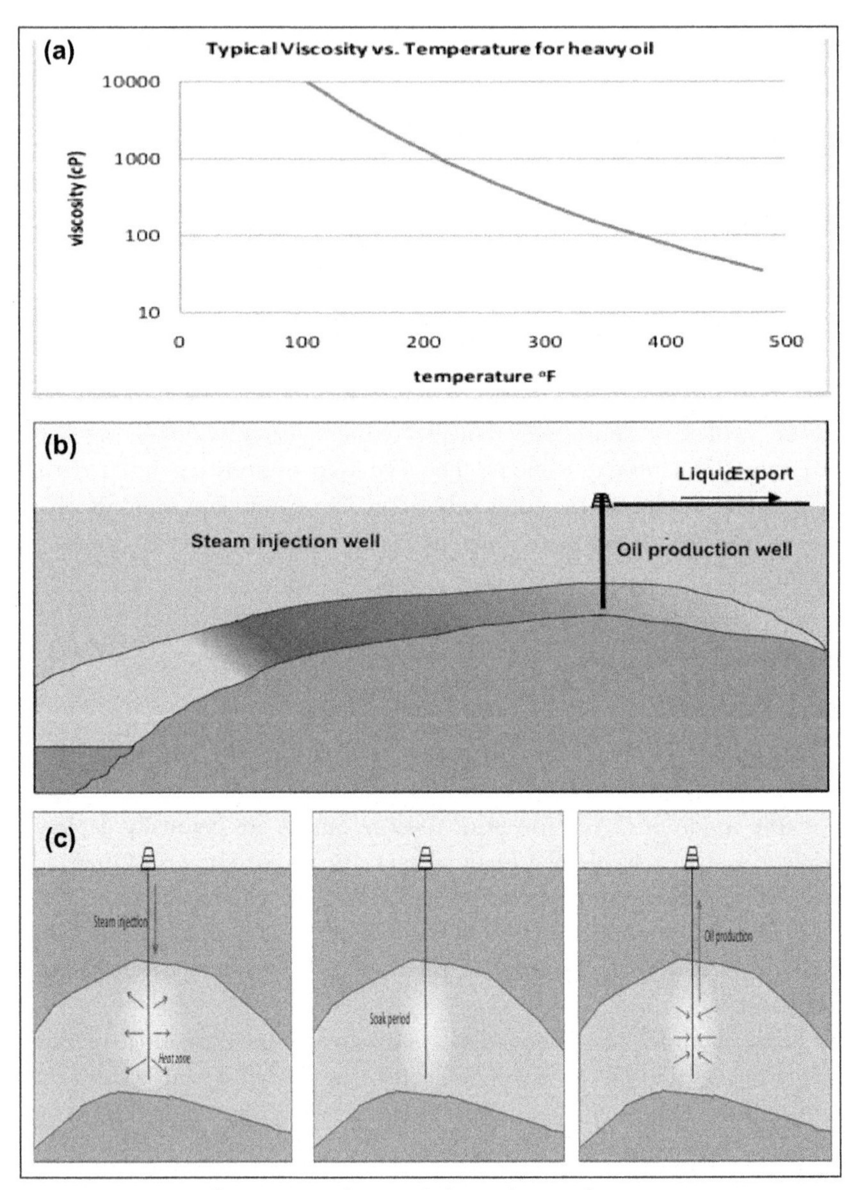

Figure 9.6 Heavy oil recovery. (a) Typical dependence of viscosity on temperature. (b) Continuous steam flooding schematic. (c) Cyclic steam injection.

Gravity effects can be used to assist in oil recovery. Recovery of up to 50% is possible.

9.5.3 Cyclic Steam Injection

This method, also known as "huff and puff," consists of three stages: injection, soaking, and production (see Fig. 9.6). Steam is injected into a well for a certain amount of time (soaking period) to allow it to heat the oil in the surrounding area, reducing oil viscosity. There is then a production period when additional pressure from injection helps to drive the hot oil to the well. Recovery factors will normally be below 20%.

9.5.4 Combustion Methods

Combustion methods involve the injection of air with subsequent ignition and combustion of oil. As the fire burns the fire front moves toward the production wells, heating the oil and reducing its viscosity. Also connate water present is vaporized—expanding and providing drive. The process is shown schematically in Fig. 9.7. It has been successful in some fields where other recovery methods are not expected to be usable.

9.5.5 Cold Heavy Oil Production with Sand

Here reservoir sand is deliberately produced to enable oil to be produced as well. Cold heavy oil production with sand exploits the finding that sand

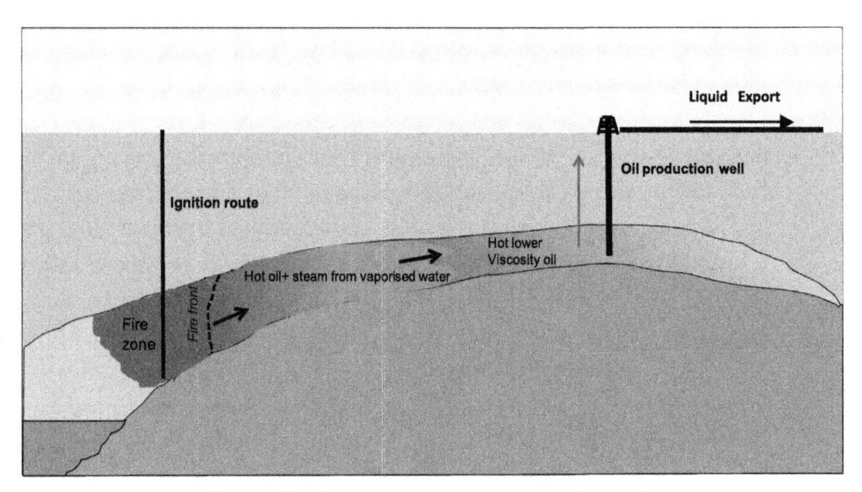

Figure 9.7 Combustion production schematic.

ingress can enhance the oil rate by an order of magnitude or more in heavy oil. The approach involves applying repeated pressure pulses to the reservoir. This has the effect of suppressing instabilities such as viscous fingering or permeability channeling, overcoming capillary barriers, and reducing pore-throat blockage.

9.6 QUESTIONS AND EXERCISES

Q9.1. List the factors that will determine the volume accessed by fracturing of shale gas.

Q9.2. What would be a typical range of RFs for shale gas?

Q9.3. Explain the meaning of HFV, SFV, and accessed volume in relation to shale gas.

Q9.4. Explain the physical nature of CBM.

Q9.5. We have a CBM play of 14,500 acres with the measured parameters shown in the table below.

Fairway area acres	Pressure psi	Net coal ft	In situ density $g/cm^3 = ton/m^3$	In situ gas content sm^3/t	Ash content %	Moisture %
14,500	497.2	88.5	1.49	3.78	0.046	0.26

If the Langmuir isotherm for the coal concerned is that above and abandonment pressure is 150 psi, estimate the RF and hence the recoverable gas from the CBM play. If 30 wells are planned, what would we expect the recovery per well to be?

Q9.6. Use the software ("gas decline-zz" and "economic indicators-zz") to determine the single-well economic viability (breakeven gas price) for a shale gas play with the following averaged properties.

Technical data: horizontal length 4500 ft, initial pressure 9000 psi, reservoir temperature 200°F, gas viscosity 0.03 cP, wellbore radius 0.25 ft, average initial production rate 12,000 mmscf/d, and average cumulative well production 3.5 bcf. Assume a reservoir porosity of 16% and a bottom hole pressure of 3000 psi. (Hint: use effective radius and permeability to match initial rate and cumulative production.)

Economic data: cost of well = \$7 mm, facilities costs = \$1 mm, discount rate = 6%, inflation = 3%, and taxation = 40%.

9.7 FURTHER READING

Y.Z. Ma, Unconventional Oil and Gas Resources Handbook, Elsevier, 2015.

J.G. Speight, Shale Gas Production Processes, Elsevier, 2013.

R. Flores, Coal and Coal Bed Methane, Elsevier, 2013.

C. Zou, et al., Unconventional Petroleum Geology, Elsevier, 2013.

V. Bakshi, Shale Gas, Global Law and Business (2012).

W. Hefley, Y. Wang, Economics of Unconventional Shale Gas Development, Springer, 2015.

CHAPTER 10

Producing Field Management

10.1 INTRODUCTION

An important part of the work of a reservoir engineer will be in the monitoring and management of producing fields. In this chapter we outline the key elements of this. The three elements in which the reservoir engineer is involved are shown in Fig. 10.1.

Because of the innate uncertainties in early reservoir models, field management can be critical to the success of a field. It is particularly important with large fields where there is significant scope for additional development stages and rectification of initial development plan errors or operational procedures. The process will be ongoing throughout the life of a field, determining its ultimate economic value.

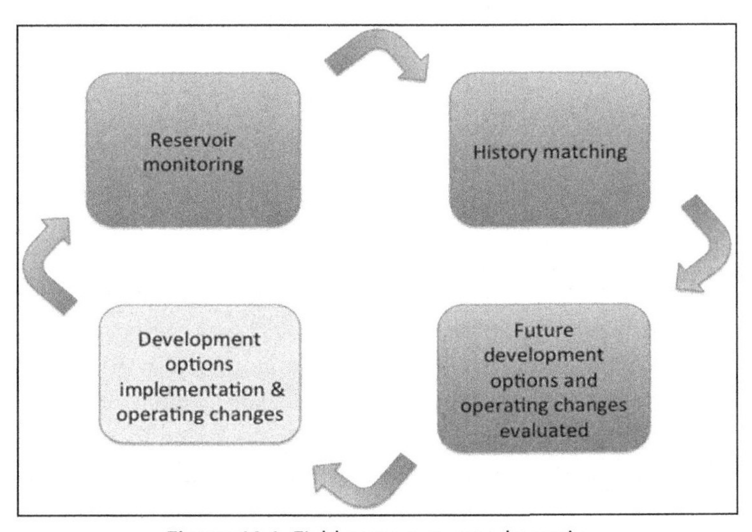

Figure 10.1 Field management schematic.

Fundamentals of Applied Reservoir Engineering
ISBN 978-0-08-101019-8
http://dx.doi.org/10.1016/B978-0-08-101019-8.00010-7
© 2016 Elsevier Ltd.
All rights reserved.

10.2 RESERVOIR MONITORING

As discussed in chapter "Numerical Simulation Methods for Predicting Reservoir Performance," once a field is producing, data starts to become available which can be used to modify and improve our reservoir model. This data comes under the following main headings.

10.2.1 Production Rates

Fluid rates (oil, gas, and water) are normally continuously monitored in real time at the wellhead, with data from individual wells sent directly to a central monitoring centre for analysis. Multiphase meters are now normal. Downhole flow meters are also available, but less commonly used as they are expensive to install and maintain and prone to breakdown.

10.2.2 Pressure Data

Flowing tubing head pressure is always available, but bottom hole pressure is less often available to the reservoir engineer. Again, downhole gauges are expensive to install and frequently break down, although the situation is improving as gauges become more robust. This is unfortunate, since the wellbore hydraulics analysis needed to translate tubing head pressure to bottom hole pressure can be difficult in multiphase systems.

Shut-in pressure data is very important for transient pressure analysis (discussed in chapter: Well-Test Analysis). Downhole shut-in is preferable if available, to avoid wellbore storage.

Repeat formation testing, discussed in Section 2.7.5 of chapter "Basic Rock and Fluid Properties" in obtaining fluid samples, will also give formation pressure profiles in wells that have been shut-in after production.

10.2.3 Tracer Data

Where we have injection of water or gas, tracer data can provide valuable information on flow patterns in the reservoir. This can be valuable in further development planning or operational management of a field. We have radioactive and chemical tracers that can be added to injected fluids and monitored for levels in surrounding producing wells.

10.2.4 Phase Saturation Distribution—4D Time Lapse

During field life, new time-dependent seismic data can be very useful in understanding the advance and developments of water fronts due to aquifer influx in injected water.

10.3 RESERVOIR HISTORY MATCHING AND REMODELING

This is the process by which a reservoir model (geological and petrophysical) is adjusted to match the production, saturation, and pressure history of the reservoir. A history-matched reservoir simulation model will more accurately predict the future performance and better represent the current pressure and saturation state of the reservoir. Actual well rates are input and, as discussed in chapter "Numerical Simulation Methods for Predicting Reservoir Performance," the following are commonly corrected in the reservoir model to obtain a better match to actual reservoir data:

- grid cell permeabilities across the reservoir
- grid cell porosities or net to gross values
- faults—their location and transmissibility
- geological extent of reservoir
- strength of aquifer.

Less commonly saturation and PVT (pressure/volume/temperature) properties such as model relative permeabilities, capillary pressure curves, and fluid properties are adjusted. The latter will be most significant with gas condensate fields, where PVT properties are particularly important.

As discussed in chapter "Numerical Simulation Methods for Predicting Reservoir Performance," it is particularly important that a geologically consistent approach is applied in history matching. A "sticking plaster" approach around individual wells should definitely be avoided!

10.4 REVIEW OF DEVELOPMENT AND MANAGEMENT OPTIONS

When a new and history-matched reservoir model is available, the reservoir engineer will need to consider how future production and economics can be enhanced. The following are some options that will need to be examined when we consider further field development.

- Adjustment to some individual well rates (or shut-in) to prevent early water or gas breakthrough.
- Further production and/or injection wells.
- Additional wells from further development phases.
- Further appraisal wells or well testing.
- Changes to facilities (such as separators in a gas condensate case) to improve efficiency.

- Stimulation of some or all wells.
- Consideration of tertiary (enhanced) recovery methods.

In all cases reservoir models will need to be constructed and optimized with economic evaluation.

10.5 FURTHER READING

N. Meeham, Reservoir Monitoring Handbook, Gulf Publishing, 2011.

J.R. Gilman, C. Ozgen, Reservoir Simulation: History Matching & Forecasting, SPE, 2013.

A. Tarek, N. Meehan, Advanced Reservoir Management and Engineering, Elsevier, 2011.

CHAPTER 11

Uncertainty and the Right to Claim Reserves

11.1 WHAT ARE RESERVES AND RESOURCES?

What a company can legitimately claim to "own" in terms of hydrocarbon reserves and resources is critically important to its perceived value and hence its share price. The stock market will value a company on this basis, and also on how rapidly it is likely to monetize these resources. (The distinction between reserves and resources will be made clear as we go on.) Realistic reporting of reserves numbers is clearly very important.

There are three factors in estimating reserves (Fig. 11.1).

1. What is in the ground?
2. What proportion of this can technically be recovered?
3. What proportion of this can be economically produced?

A broad definition of reserves is therefore: reserves are estimated quantities of hydrocarbons which can be technically and economically recovered from a known accumulation at a given point in time.

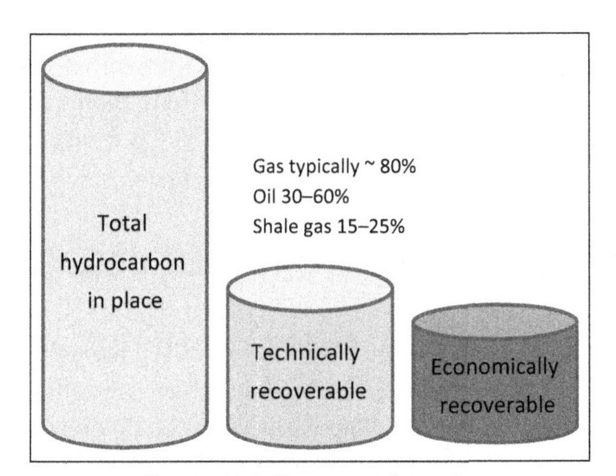

Figure 11.1 Reserves schematic.

Fundamentals of Applied Reservoir Engineering
ISBN 978-0-08-101019-8
http://dx.doi.org/10.1016/B978-0-08-101019-8.00011-9
© 2016 Elsevier Ltd.
All rights reserved.

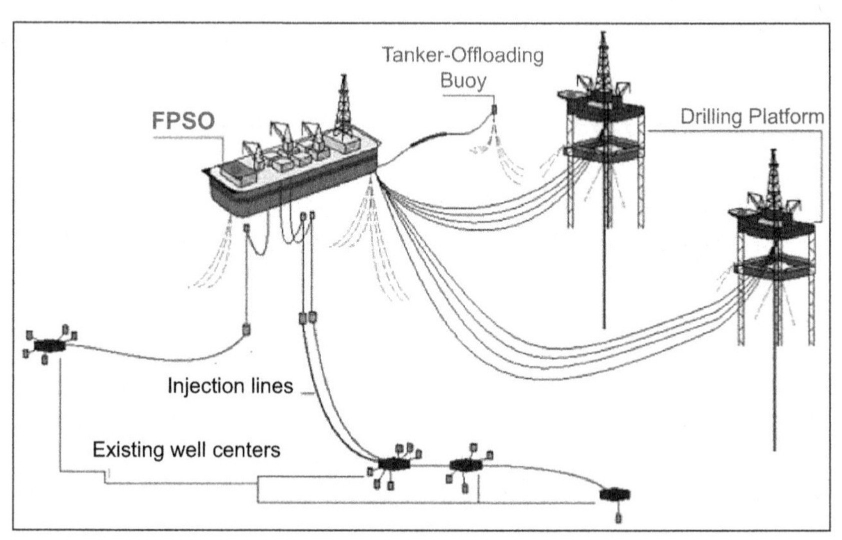

Figure 11.2 Example development.

To support reserves or resource estimates, the appropriate level of economic and commercial maturity is required. This implies a reasonable expectation of a market, transport to that market, and positive economics.

We can have no reserves for a field if there is no economic development. For example, Fig. 11.2 shows a potential oil-field development, but to claim reserves a company needs to show that there is an economic development plan to extract them.

A company's responsibility to investors and shareholders requires a balanced and auditable approach to reserves estimation and reporting.

Underbooking of reserves gives an incorrect picture of company share value, and adversely affects corporate metrics and therefore share price. *Overbooking* carries reputational risk and involves governance issues, and hence would have serious share price consequences.

11.2 INTERNATIONAL RULES ON PUBLIC DECLARATION OF RESERVES

The US Securities and Exchange Commission (SEC) has published rules on how classes of reserves should be estimated and declared (they concentrate very much on proved reserves). These rules must be followed by companies whose shares are quoted on the New York Stock Exchange. Also the Society of Petroleum Engineers has guidelines that in fact now differ very little from SEC rules.

11.3 HANDLING UNCERTAINTIES ON RESERVES

11.3.1 Uncertainty Overview

Reserves uncertainty arises from:

commercial uncertainty—political, market, and transportation
technical uncertainty—geological and engineering
economic uncertainty—future gas and oil prices, and development and operating costs.

All three of these may be expected to decrease as a project matures, with appraisal and development planning and optimization, and later when production starts. A useful way of representing this is shown in Fig. 11.3.

This range is normally defined in probabilistic terms, with a P10 upside, a P50 best estimate, and a P90 downside. These are mathematically defined as follows.

1. **P10** probability case: there is a 10% chance of reserves being at this level or more than this, and a 90% chance of reserves being less = upside or high-side estimate.
2. **P50** probability case: there is a 50% chance of the reserves being less than this and a 50% chance of reserves being more = best estimate.

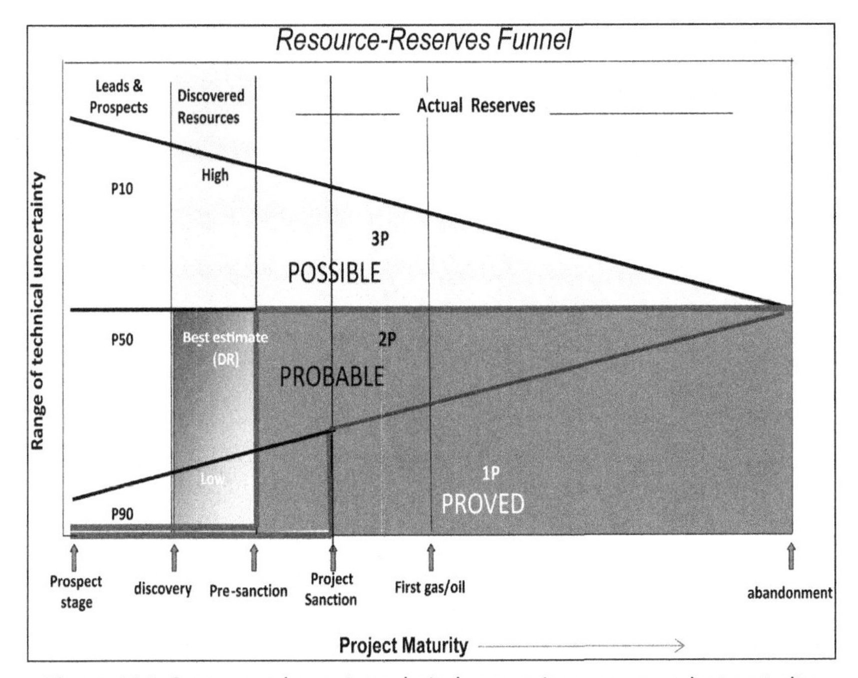

Figure 11.3 Reserves schematic: technical uncertainty versus project maturity.

3. **P90** probability case: there is a 90% probability of there being more reserves than this and a 10% chance of reserves being at this level or less.

With increasing project maturity, uncertainty decreases and P10, P50, and P90 converge as shown in Fig. 11.3, until they are equal at field abandonment.

There are two ways in which reserves uncertainty is defined and estimated—deterministic and probabilistic.

11.3.2 Deterministic Estimation of Reserves

Nomenclature:

proved reserves (1P) = modeled with single set of parameters
proved + probable (2P) = single set of "best estimate" parameters
proved + probable + possible (3P) = single set of parameters.

From the results of a **sensitivity analysis**, as discussed in chapter "Numerical Simulation Methods for Predicting Reservoir Performance," the reservoir engineer must choose a combination of the major uncertainty parameters that represent **realistic** downside and upside cases.

The engineer must take care not to multiply downsides in the 1P case or upsides in the 3P case. Take an example where we have three major uncertainties, as shown in Fig. 11.4, with best estimate (2P) reserves of 100 mmbbl.

A reasonable downside **proved reserves** case would be to take the "fault uncertainty" P90 case (−34 mmbbl), leaving all other parameters at their P50 levels, so that downside reserves would be 100 − 34 = 66 mmbbl. It would **not** be reasonable to allow for downside "petrophysics," "heterogeneity," etc. as well—it is not realistic to assume that all of the "bad" downside cases

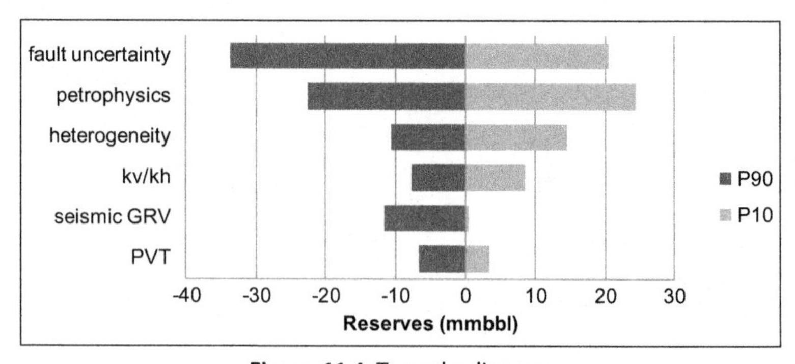

Figure 11.4 Tornado diagram.

occur together! This would in fact be $(1 - 0.1 \times 0.1 \times 0.1) = (1 - 0.001) =$ a P99.9 probability case.

As an upside **proved + probable + possible** case, the P10 "petrophysics" case (+25 mmbbl) could be used, giving 3P (proved + probable + possible) = P10 reserves of 100 + 25 = 125 mmbbl.

We thus have the reserves range:

proved reserves (1P) = 66 mmbbl

proved + probable reserves (2P) = 100 mmbbl

proved + probable + possible reserves (3P) = 125 mmbbl.

11.3.3 Probabilistic Estimation of Reserves

The nomenclature corresponds to the above definitions, with a P90 − P50 − P10 range. Proved reserves are assumed to be 1P ~ P90, 2P ~ P50, and 3P ~ P10.

11.3.3.1 Monte Carlo Analysis

A Monte Carlo analysis can be used to calculate the range of probability for reserves. The parameters need to be independent (or interdependent in a relatively simple way):

$$\text{Reserves} = \text{Gross Rock Volume (GRV)} \times \text{fraction containing hydrocarbons } (F) \times \text{expansion/shrinkage factors } (E) \times \text{recovery factor (RF)} \quad [11.1]$$

Therefore there are four factors to be considered—GRV, *F*, *E*, and RF—with reserves being the product of these. Monte Carlo analysis randomly selects combinations of these parameter values (according to the distributions that we give it) using a computer algorithm many times, and builds up a reserves distribution curve like that shown in Fig. 11.5, where the vertical axis will depend on the number of times that each reserves number occurs.

Monte Carlo is more appropriate for exploration-stage fields than during the development stage, where a set of "deterministic" (ie, scenario-based) reservoir models must be built.

11.3.3.2 Experimental Design

Experimental design enables Monte Carlo methods to be used where we have an existing numerical simulation model.

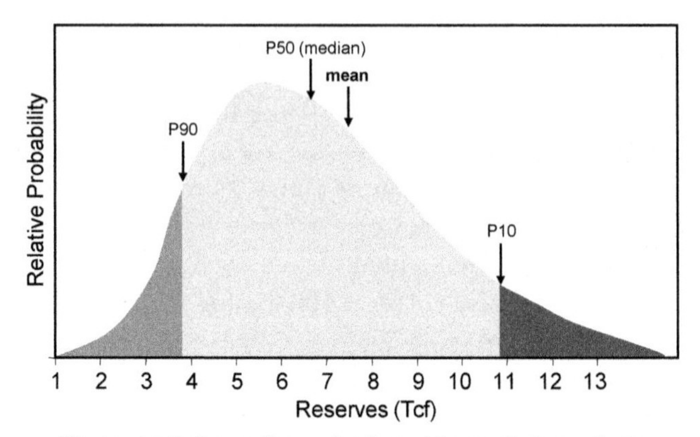

Figure 11.5 Example results from Monte Carlo analysis.

A polynomial surface is generated. The example in Fig. 11.6 shows a polynomial surface for a two-parameter case. Where the P90, P50, or P10 surfaces cut the polynomial surface, all points on the intersection line give reserves at these respective levels.

We need to choose one combination of parameters from the P90, P50, and P10 surfaces as the basis for each of the 1P, 2P, and 3P deterministic models.

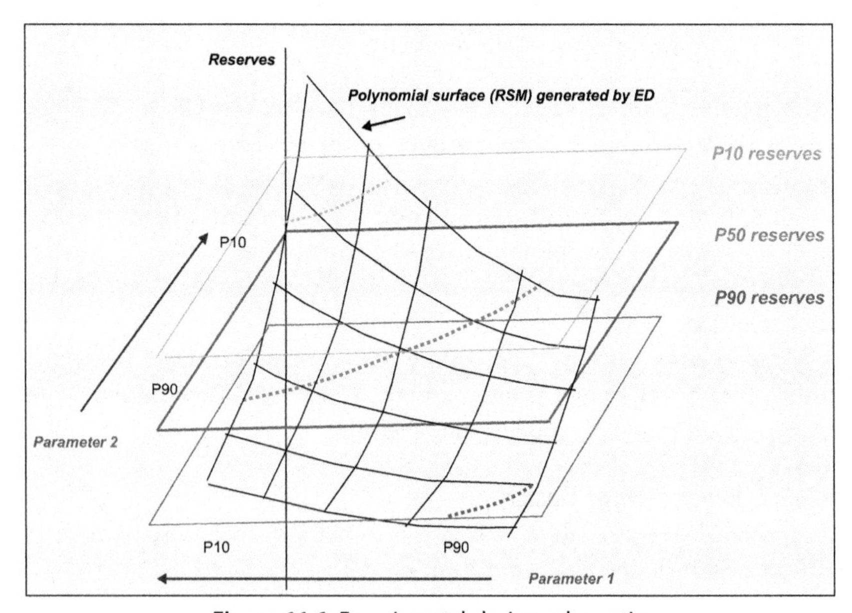

Figure 11.6 Experimental design schematic.

11.4 PUBLIC DECLARATION OF RESERVES

11.4.1 Prospect Stage Resources

In the **prospect stage** we have the greatest uncertainty. On discovery we have **discovered resources**.

11.4.2 Discovery to Presanction Resources

In the stage between discovery and presanction we have resources, but no reserves defined as yet. We have what are normally described as "discovered resources" or sometimes as "contingent resources."

11.4.3 Presanction to Project Sanction Reserves

There is a presanction point where the following apply.

1. Project sanction is expected within three to five years (different for different companies).
2. We have positive market conditions—the demand/supply position looks favorable.
3. There are clearly identifiable markets and transportation.
4. There is a development plan with timescales, platform/well numbers, costs, etc. (if nonoperated this would normally be the operator's plan, even if this is rather general).
5. We have a technical model corresponding to this, meeting the normal criteria with a P50 profile.
6. Reasonable assumptions can be made on host-government approval and partner alignment.

Probable reserves are now normally declared by companies: these correspond to their P50 estimates.

11.4.4 Project Sanction to First Oil or Gas Reserves

Project sanction is when a final investment decision is made and a company is financially committed to a development.

During this stage facilities are constructed and wells drilled.

Proved reserves are declared publicly at this stage. These will correspond to the P90 estimate. Also 2P (proved + probable reserves), updated as necessary, are declared.

11.4.5 Fields in Production Reserves

Proved and proved + probable reserves, revised annually, continue to be declared publicly during the life of the field.

At any stage companies do not necessarily declare reserves for individual fields, but often only for regional interests.

11.5 QUESTIONS AND EXERCISES

Q11.1. Explain the relationship between total in-place resources, technically recoverable resources, and economically recoverable resources.

Q11.2. The table below shows results from a sensitivity analysis on an oil field at the project sanction stage and a best estimate of 100 mmbbl reserves. Use the software provided to obtain a tornado diagram based on the data below. Suggest a range of reserves (1P, 2P, 3P) for public release.

Base case reserves (mmbbl)		100
	P90	**P10**
Seismic GRV	56	115
Petrophysics	87	125
Faults	95	110
k_v/k_h	92	108
Heterogeneity	88	100
PVT	93	103

Q11.3. Explain Monte Carlo analysis.

11.6 FURTHER READING

Guidelines for application of the petroleum resource management system (PRMS), SPE, 2007 (online).

R. Wheaton, C. Coll, Reserves Estimations Under New Sec 2009 When Using Probabilistic Methods, SPE131241, 2009.

Centre for Economics and Management (IFP School), Oil and Gas Exploration and Production (Reserves, Costs, Contracts), Editions Technip, 2007.

APPENDIX 1

Basic Fluid Thermodynamics

A system may undergo a spontaneous change for one or both of two reasons.

1. To minimize energy.
2. To maximize entropy.

Gibbs free energy is a measure of both of these, and for any change occurring at a constant external pressure p the change in free energy (ΔG) tells us whether the change will occur spontaneously or not (Fig. A1.1). Now:

$$\Delta G = \Delta E + p\Delta V - T\Delta S \qquad \text{[A1.1]}$$

where ΔG = change in Gibbs free energy; ΔE = change in internal energy; ΔV = change in volume; and ΔS = change in entropy.

GIBBS FREE ENERGY OF MIXING

If ΔG is negative (due to negative ΔE or positive ΔS, or both) the change will occur spontaneously.

Consider the mixing of a number of components.

Looking at a simple two-component system:

$$\Delta G^{\text{mixing}} = \Delta E^{\text{mixing}} + p\Delta V - T\Delta S^{\text{mixing}} \qquad \text{[A1.2]}$$

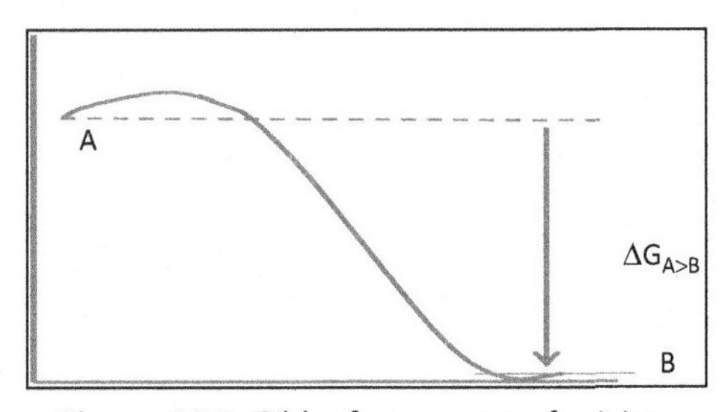

Figure A1.1 Gibbs free energy of mixing.

ENTROPY OF MIXING

ΔS^{mixing} is always positive.

Entropy depends on the number of possible "arrangements" of molecules, and in mixtures it is always more than in separate phases, therefore mixing will always occur from the entropy effect due to the negative third term in the equation for ΔG^{mixing}. The entropy change of mixing is shown in Fig. A1.2b for a two-component mixture.

INTERNAL ENERGY OF MIXING

For a two-component mixture, if the attractive interaction between types 1 and 2 molecules is less than the average between types 1 and 1 and 2 and 2, then **ΔE^{mixing} will be positive for all mixtures** (Fig. A1.2b).

GIBBS FREE ENERGY OF MIXING—COMBINATION OF TERMS

$$\Delta G^{\text{mixing}} = \Delta E^{\text{mixing}} + p\Delta V - T\Delta S^{\text{mixing}} \qquad \text{[A1.3]}$$

Combining entropy and internal energy effects will thus reduce free energy by splitting into coexisting phases α and β, as shown in Fig. A1.2c.

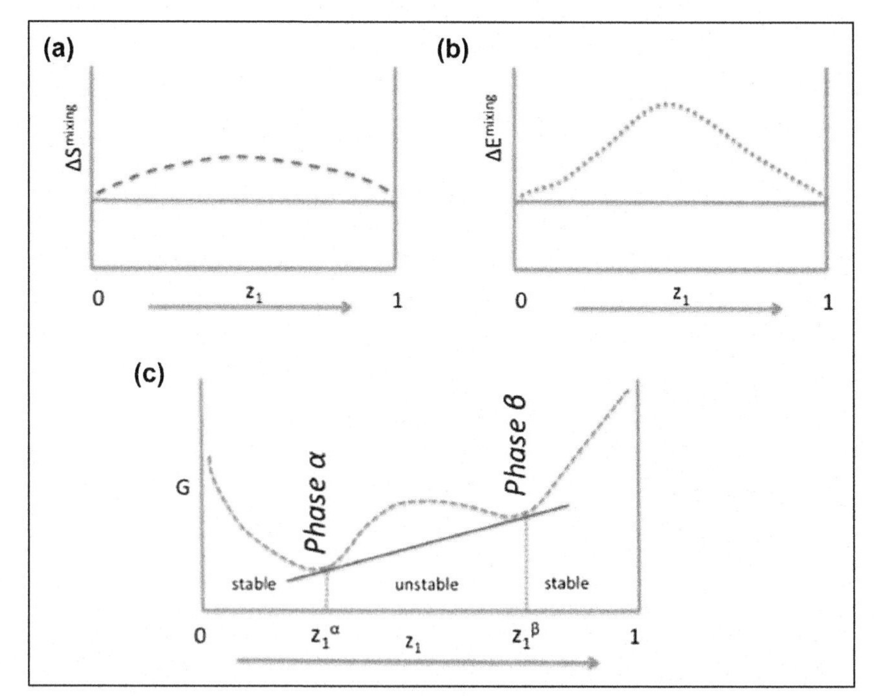

Figure A1.2 Gibbs free energy of a two-component mixture. (a) Entropy of mixing; (b) Internal energy of mixing; (c) Total Gibbs free energy of a two-component mixture.

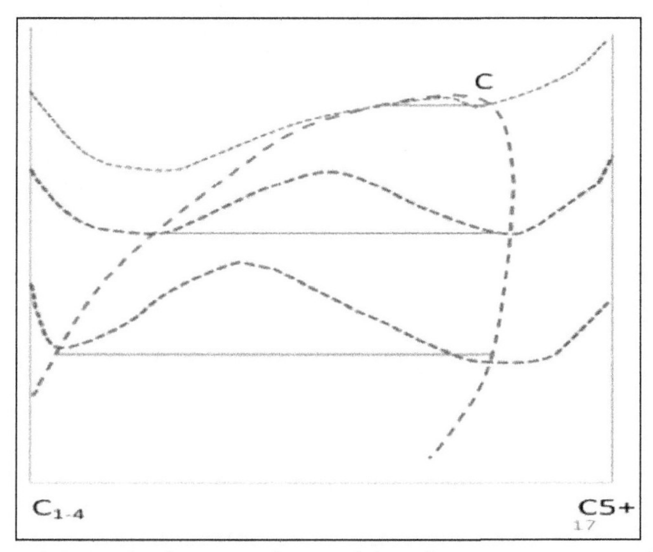

Figure A1.3 Relationship between phase stability of a two-component mixture and composition and pressure.

A mixture of components 1 and 2 with composition (z_1,z_2) between $z_1{}^{\alpha}$ and $z_1{}^{\beta}$ will be unstable, and will lose free energy by splitting into phases α and β with compositions $\left(z_1{}^{\alpha}, z_1{}^{\alpha}\right)$ and $\left(z_1{}^{\beta}, z_1{}^{\beta}\right)$ respectively. Mixtures whose compositions lie in the regions $0 < z_1 < z_1{}^{\alpha}$ and $z_1{}^{\beta} < z_1 < 1$ will be thermodynamically stable and will not split into separate phases.

Changes in pressure and temperature of systems will alter the arrangement of molecules, both in the case of pure components and for mixtures. Because of this, $\Delta E^{\mathrm{mixing}}$ and thus free energy as a function of (z_1, z_2) will change with pressure and temperature.

PHASE SPLIT FOR TWO-COMPONENT MIXTURES

For a set of different pressures the variation in Gibbs free energy of mixing (indicated by red broken lines in Fig. A1.3) shows the origin of the two-phase envelopes we saw with hydrocarbon mixtures.

APPENDIX 2

Mathematical Note

The use of mathematical calculus has been deliberately kept to a minimum in this text and where it is necessary, in most cases an attempt is made to explain the meaning in the text. Such equations are not popular with some students, but they are the basis of the behavior of reservoirs and are a concise representation of the physical relationships involved. It is therefore worth having a mathematical note here where an attempt is made to clarify the significance of the various mathematical operators used in this and other texts.

GRADIENTS

Where we have pressure (p) as a function of a single parameter, normally distance (x), the gradient (or slope) *at a point* x (ie, in the limit Δx goes to 0) is $= \frac{dp}{dx}$, as shown in the plot in Fig. A2.1. This is the case in Darcy's law, where flow rate is proportional to the rate of change (gradient) of pressure with distance through the porous media.

Where we have pressure as a function of a number of parameters, normally distance and time (x and t) or directional distances and time (x,y,z), we have what are called "partial" derivatives, as in the example in Fig. A2.2, where pressure as a function of both distance and time is represented by a two-dimensional surface, and the gradients with distance (x) **keeping time (t) constant** at a point (x,t) is represented by $\left(\frac{\partial p}{\partial x}\right)_t$. Similarly, the gradient

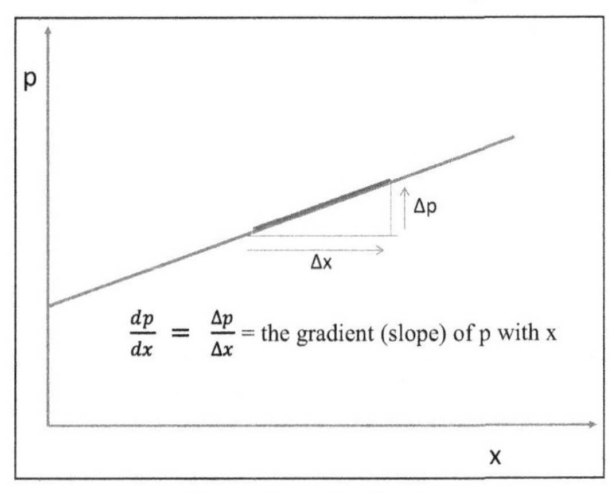

Figure A2.1 Gradients.

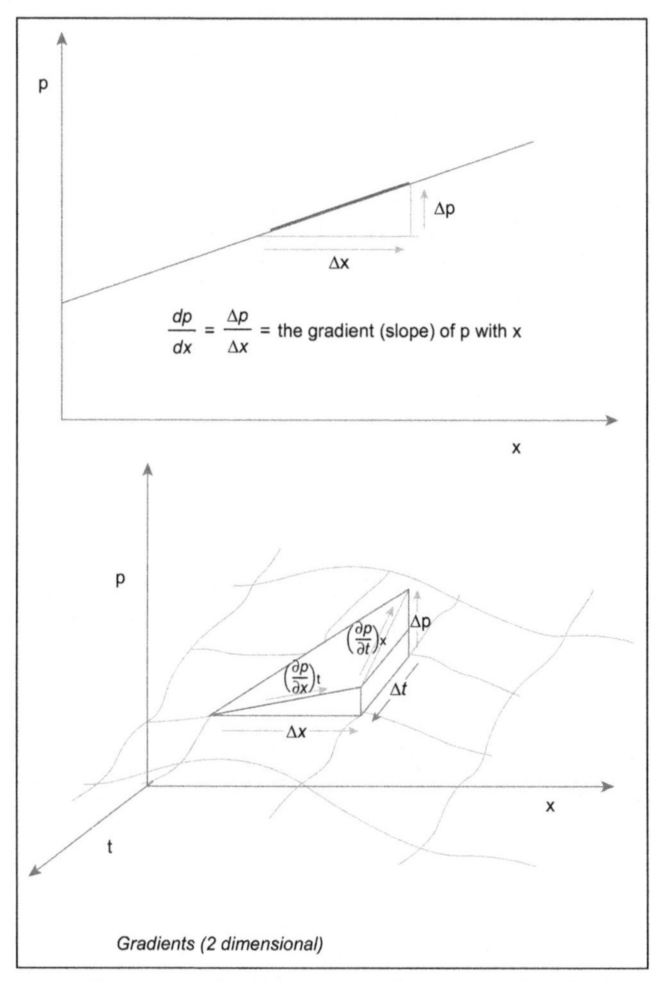

Figure A2.2 Gradients (one and two dimensional).

with time keeping x constant is represented by $\left(\frac{\partial p}{\partial t}\right)_x$: $\left(\frac{\partial p}{\partial t}\right)_r$, normally written just as $\frac{\partial p}{\partial t}$. This is seen in well-test analysis equations, along with the derivative of density $\frac{\partial \rho}{\partial t}$.

When we consider the Darcy law in a three-dimensional context, we use what is known as the "grad" operator which allows for directional components:

$$\nabla = \frac{\partial}{\partial x}\boldsymbol{i} + \frac{\partial}{\partial y}\boldsymbol{j} + \frac{\partial}{\partial z}\boldsymbol{k}$$

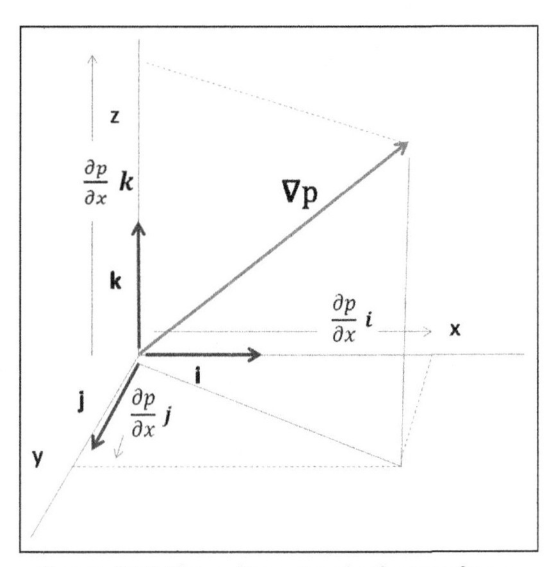

Figure A2.3 Three-dimensional reference frame.

so that in the Darcy law:

$$\nabla p = \frac{\partial p}{\partial x}\boldsymbol{i} + \frac{\partial p}{\partial y}\boldsymbol{j} + \frac{\partial p}{\partial z}\boldsymbol{k}$$

is now a vector with directional components in the x, y, and z directions, as shown in Fig. A2.3.

So that flow velocity (for a horizontal system) is a vector **with directional components** where:

$$\boldsymbol{u} = -\frac{k}{\mu}\nabla p$$

or:

$$\boldsymbol{u} = -\frac{k}{\mu}\left(\frac{\partial p}{\partial x}\boldsymbol{i} + \frac{\partial p}{\partial y} + \boldsymbol{j}\frac{\partial p}{\partial z}\boldsymbol{k}\right)$$

In the conservation of mass equation:

$$\nabla.(\rho\boldsymbol{u}) + Q_{\text{well}} = -\frac{\partial(\varnothing\rho)}{\partial t}$$

we have what is called the "divergence" operator (**∇**.). This operates on a vector, so that $\nabla.\boldsymbol{a} = \left(\frac{\partial a_x}{\partial x} + \frac{\partial a_y}{\partial x} + \frac{\partial a_x}{\partial x}\right)$. We therefore have a term that represents the total movement of mass into our volume element from the three directions (x, y, and z).

APPENDIX 3

Gas Well Testing

Solving the basic equations in a case where we cannot assume an incompressible fluid, we obtain for a gas:

$$m(p_w) = m(p_i) - \frac{qT}{kh} \cdot \left\{ \ln\left(\frac{kt}{\varphi \mu c r_w^2}\right) + 0.80907 \right\} \qquad [A3.1]$$

where $m(p)$ is the pseudopressure function replacing actual pressure:

$$m(p) = 2\int_{p_b}^{p} \frac{pdp}{\mu Z} \qquad [A3.2]$$

Here p_b is a base pressure, so that:

$$m(p_w) - m(p_i) = 2\int_{p_b}^{p} \frac{pdp}{\mu Z} \qquad [A3.3]$$

where $p_w(t)$ = pressure at the wellbore at time t; p_i = the initial pressure; Z = gas compressibility factor; and p_w = wellbore pressure, so the pressure change for gases is a function of the dependence of $(p/\mu Z)$ on pressure.

For a typical gas the relationship is like that shown in Fig. A3.1.

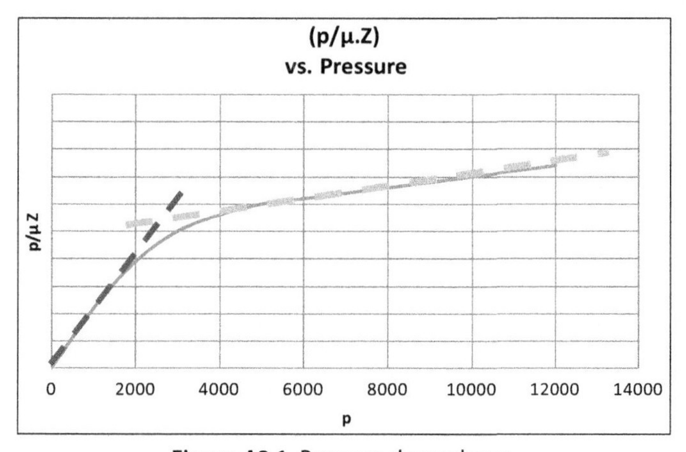

Figure A3.1 Pressure dependence.

There are therefore large pressure regions where $\frac{p}{\mu Z}$ is essentially linear with pressure. If our upper and lower pressures are in one of these ranges, the gradient will be the average $(p - p_w)/2\,\mu Z$, so that

$$\int_{p_i}^{p_w} \frac{pdp}{\mu Z} = \frac{1}{2\mu Z}(p + p_w)\int_{p_w}^{p} pdp = \frac{1}{2\mu Z}\left(p^2 - p_w^2\right) \qquad [A3.4]$$

This is a commonly used simplification, and gives us the transient equation in field units:

$$p(r,t)^2 = p_i^2 - \frac{1{,}637 \cdot qZT\mu}{kh} \cdot \left\{ \log\left(\frac{kt}{\varphi\mu c r^2}\right) - 3.23 \right\} \qquad [A3.5]$$

Another possible simplification for higher pressures is to assume that $\frac{p}{\mu Z}$ = a constant over the range considered, in which case

$$\int_{p_i}^{p_w} \frac{pdp}{\mu Z} = \frac{p}{\mu Z}(p_i - p_w) \qquad [A3.6]$$

which give an equation equivalent to the oil transient equation.

APPENDIX 4

Enhanced Oil Recovery

GENERAL

We have seen that recovery factors in oil reservoirs can be between 20% and 60%—the higher figure normally depending on water flooding (known as secondary recovery). Thus 40—80% of the oil in place can be left behind, mainly due to high oil viscosities (heavier oils), high residual oil saturation (a function of oil—water—solid interfacial forces), and poor areal sweep efficiency. A number of enhanced recovery methods, known as tertiary recovery methods, can help to overcome these problems.

GAS INJECTION

Injected gas can be produced gas, either processed or unprocessed, or gases such as CO_2, nitrogen, or mixtures of produced gas and CO_2.

Gas injection or water-alternating gas injection is the most widely used enhanced oil recovery method. Like water injection, gas injection keeps reservoir pressure higher, which will on its own increase deliverability. Sweep efficiency can be improved, particularly in high-relief reservoirs where gravity drainage can be significant—with gas injected higher in the reservoir the gravity effect (oil is heavier than gas) can help drive oil to lower production wells. Water-alternating gas can give more control on sweep of oil toward producers. Swelling of the oil and vaporization of oil components can both help recovery.

Gas injection is classified as either miscible or immiscible. Miscible gas injection (where the gas mixes or partially mixes with the oil) can reduce oil viscosity, reduce oil/gas interfacial tension, and change wetting properties such that residual oil saturation is reduced.

MISCIBLE SOLVENTS—SURFACTANTS

Injection of surfactants can reduce oil/water interfacial tension and thus reduce residual oil saturation and increase oil recovery.

The surfactant flooding technique normally uses separate injection and production wells to increase oil recovery. Improvement occurs by reducing both interfacial tension and capillary forces in the formation, increasing contact angle, and decreasing residual oil saturation.

Surface active agents are amphiphilic organic compounds with a chemical structure that consists of two different molecular components, known as hydrophilic and hydrophobic groups. These distribute themselves between the oil and water phases and reduce interfacial tension. Where oil is the wetting phase, the contact angle increases and we have a system where the flow of water displaces the oil droplets. This displaced oil can then move toward the production wells.

THERMAL METHODS

There are two thermal enhanced oil recovery methods: steam flooding and fire flooding.

Steam flooding heats the oil, reducing its viscosity and vaporizing part of the oil, and thus decreases the mobility ratio.

Fire flooding involves the injection of air, with subsequent ignition and combustion. As the fire burns, the fire front moves towards the production wells, heating the oil and reducing its viscosity.

ECONOMICS

All these methods add to the costs of producing the oil and oil prices, and the cost of injectants such as surfactants or CO_2 will need to be taken into account when looking at the feasibility of any enhanced oil recovery project. Numerical simulators are available which cover all of the above methods.

FURTHER READING

L. Lake, R. Johns, B. Rossen, G. Pope, Fundamentals of Enhanced Oil Recovery, SPE, 2014.

APPENDIX 5

Simple Oil Material Balance for Rate as a Function of Time

Determining production q as a function of time requires the Darcy equation:

$$q = \frac{kh(p - p_w)}{141.2\mu \cdot \ln\left(\frac{r_e}{r_w}\right)}$$

and material balance. If we neglect gas or water ingress and rock–water expansion effects, material balance gives:

$$\Delta N = N[(B_o - B_{oi}) + (R_{si} - R_s) \cdot B_g]$$

Let us assume simple linear relationships, so that:

where $p > p_b$ $\quad B_o = m_1 \cdot (p_b - p) + B_o(p_b)$

$$R_s = R_{si}$$

where $p > p_b$ $\quad B_o = m_2 \cdot (p - p_b) + B_o(p_b)$

$$R_s = m_3 \cdot (p - p_b) + R_{si}$$

for gas $\quad B_g = n_1/p = 5.044/(\mathrm{PZT})$

where $p > p_b$:

$$\sum (q \cdot \Delta t) = \Delta N = N \cdot (B_o - B_{oi})$$
$$= N \cdot [\{m_1(p_b - p) + B_o(p_b)\} - B_{oi}]$$

$$N \cdot m_1(p_b - p) = \sum (q \cdot \Delta t) - N \cdot B_o(p_b) + N \cdot B_{oi}$$

$$N \cdot m_1 \cdot p = N \cdot m_1 \cdot p_b - \sum (q \cdot \Delta t) + N \cdot B_o(p_b) - N \cdot B_{oi}$$

$$p = \left(N \cdot m_1 \cdot p_b - \sum (q \cdot \Delta t) + N \cdot B_o(p_b) - N \cdot B_{oi}\right) \Big/ N \cdot m_1$$

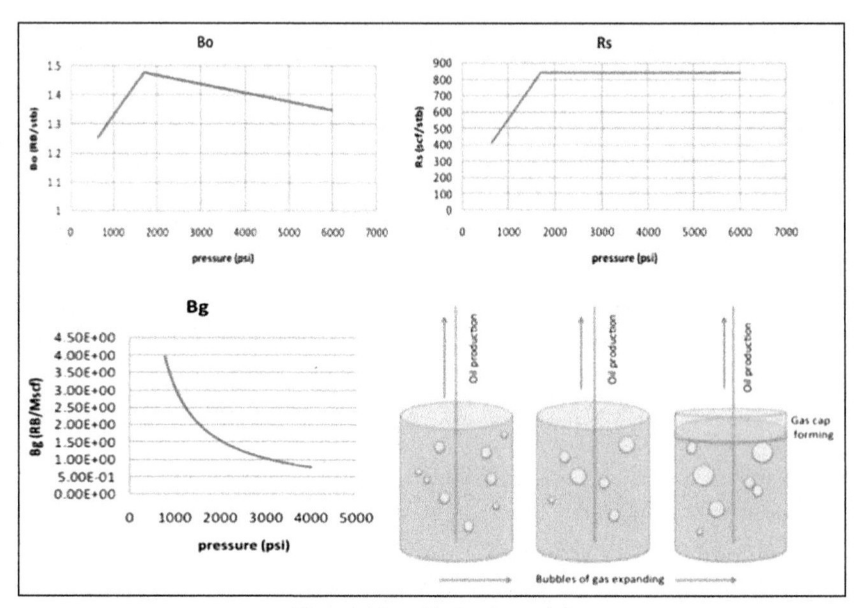

Figure A5.1 Black oil model.

and where $p < p_b$:

$$\sum(q \cdot \Delta t) = \Delta N = N \cdot [(B_o - B_{oi}) + (R_{si} - R_s) \cdot B_g]$$

$$\sum(q \cdot \Delta t) = N \cdot [(m_2(p - p_b) + B_o(p_b)) - B_{oi} + \{R_{si} - m_3(p - p_b) - R_{si}\} \cdot n_1/p]$$

$$\sum(q \cdot \Delta t) = N \cdot [(m_2(p - p_b) + B_o(p_b)) - B_{oi} - m_3(p - p_b) \cdot n_1/p]$$

$$p \cdot \sum(q \cdot \Delta t)/N = m_2 p^2 - m_2 p_b \cdot p + (B_o(p_b) - B_{oi}) \cdot p - m_3 \cdot n_1 \cdot p + m_3 \cdot n_1 \cdot p_b$$

therefore:

$$m_2 p^2 + \left[B_o(p_b) - B_{oi} - m_3 \cdot n_1 + m_2 \cdot p_b - \sum(q \cdot \Delta t)/N\right] \cdot p + m_3 \cdot n_1 \cdot p_b = 0$$

This can be written as:

$$ap^2 + bp + c = 0$$

where:

$$a = m_2 \quad b = \left[B_o(p_b) - B_{oi} - m_3 \cdot n_1 + m_2 \cdot p_b - \sum(q \cdot \Delta t)/N\right]$$

$$c = m_3 \cdot n_1 \cdot p_b$$

and

$$p = \frac{-b + \sqrt{b^2 - 4ac}}{2a}$$

We can therefore solve $p(t + \Delta t)$ as a function of $\sum(q(t) \cdot \Delta t)$ and the various black oil parameters at the pressure $p(t)$.

Solution gas drive assumes all produced gas remains within the reservoir, either as expanding bubbles of gas or by creating a gas cap (Fig. A5.1).

APPENDIX 6

Conversion Factors

Parameter	Field unit	SI unit	Conversion field to SI
Length	Foot	M	0.3048
Area	ft^2	m^2	0.0920304
	Acre	km^2	4.046873×10^{-3}
Volume	Barrel (bbl)	m^3	1.589873×10^{-1}
	acre ft	m^3	1.233482×10^{3}
	ft^3	m^3	2.831685×10^{-2}
Mass	lb mass	Kg	4.535924×10^{-1}
Temperature gradient	°F/ft	°K/m	1.822689
Pressure	Psi	Bar	0.06894757
Density	lb/ft^3	kg/m^3	1.601848×10^{-1}
Viscosity	cP	Pa·s	1.0×10^{-3}
Permeability	mD	μm^2	9.869233×10^{-4}
Velocity	$bbl/day/ft^2$	m/sec	1.9994×10^{-5}

APPENDIX 7

Answers to Questions and Exercises

Answers to definition/explanation-type questions can be found by reference to the main text. Answers to numerical questions are given in sections below the questions.

BASIC ROCK AND FLUID PROPERTIES

Q2.9. Since from the above:

$$Q = \frac{kA(P_1 - P_2)}{\mu x}$$

$$k = \mu \cdot x \cdot \left(\frac{Q}{A(P_1 - P_2)} \right)$$

now $\mu = 2.0$ cP, $x = 10$ cm, $A = 12.5$ cm^2, $(P_1 - P_2) = 50/14.7$ atm, and $Q = 0.05$ cm^3/s:

$$k = 2.0 \times 10 \times 0.05/(12.5 \times [50/14.7]) = 23.5 \text{ mD}$$

Q2.10. Since from the above:

$$Q = \frac{kA\left(P_1^2 - P_2^2\right)}{2\mu x}$$

$$k = 2\mu \cdot x \cdot \left(\frac{Q}{A\left(P_1^2 - P_2^2\right)} \right)$$

$$k = 2 \times 0.0178 \times 8.0 \times 23.6/(9.14 \times 12.5) = 145 \text{ mD}$$

WELL-TEST ANALYSIS

Q3.3. From the spreadsheet,

$$\text{Permeability,} \quad k = 66 \text{ mD.}$$

Q3.4. From the spreadsheet,

$$\text{Permeability,} \quad k = 25 \text{ mD.}$$

ANALYTICAL METHODS

Q4.1.

$$\Delta N = [N \cdot \{(B_o - B_{oi}) + (R_{si} - R_s) \cdot B_g + \Delta p \cdot B_{oi} \cdot (c_w S_{wi} + c_f)/(1 - S_{wi})\} + W_f]/\{B_o + (R_p - R_s) \cdot B_g\}$$

where $N =$ stb of oil initially in place; $\Delta N =$ stb of oil produced; $B_{oi} =$ initial fvf (RB/stb); $B_o =$ fvf at lower pressure (RB/stb); $R_{si} =$ initial solution GOR (scf/stb); $R_s =$ GOR at some other pressure (scf/stb); $R_p =$ cumulative produced GOR (scf/stb); $B_g =$ gas fvf (RB/scf); $N =$ 200 stb; B_o (800psi) (RB/stb) $= 1.278$; $B_{oi} = 1.467$ (RB/stb); $R_{si} = 838.5$ scf/stb; $R_s = 464$ scf/stb; $B_g = 0.004$ RB/scf; $R_p = 800$ scf/stb. Dimensions: term one RB/stb; term two scf/stb × RB/scf = RB/stb; within {} = RB; denominator RB/stb; $(R_p - R_s)$; and $B_g =$ scf/stb × RB/scf.

$$\begin{aligned}\Delta N &= 200 \times \{(1.278 - 1.467) + (838 - 464) \times 0.004\}/(1.278 \\ &\quad + (800 - 464) \times 0.004) \\ &= (-37.8 + 299.2)/2.62 = 99.7 \text{ stb}\end{aligned}$$

Recovery factor $= \Delta N/N = 99.7/200 = 50\%$.

Q4.2.

$$\Delta V^o = V_i^o \left(1 - \frac{pZ_i}{Zp_i}\right)$$

$\Delta V^o =$ volume of gas produced (bscf—surface conditions); $V_i^o =$ gas initially in place (bscf—surface conditions); $p_i =$ initial pressure (psi); $p =$ final pressure (psi); $Z_i =$ compressibility at initial conditions ($Z_i = f(p_i)$); and $Z =$ compressibility under final conditions ($Z = f(p)$):

$$\Delta V^o = 200 \times (1 - 1000 \times 0.86/3000 \times 0.8) = 128 \text{ bscf}$$

Q4.3. Water saturation at breakthrough = 53%.
Water saturation behind breakthrough front = 65%.
Recovery factor at breakthrough = (65 − 20) = 45%.

Q4.5. Water saturation at breakthrough = 45%.
Water saturation behind breakthrough front = 55%.

Q4.7. Reserves = 16.94 bcf; RF = 69%.

Q4.8. Gas reserves = 16.94 bcf; RF = 69%.
Liquid reserves = 1.41 mmstb; RF = 69%.

Q4.9. Recoverable oil = 19.75 mmstb.
Recoverable gas = 11.77 bcf.
Oil recovery factor = 55.7%.

ESTIMATION OF RESERVES

Q6.1.

$$N = 7758 A h_v \varphi (1 - S_w)/B_{oi}$$

$$A = 2400 \text{ acres}$$

$$h_v = (0.9 \times 200)$$

$$\phi = 0.15$$

$$S_w = 0.20$$

$$B_{oi} = 1.48 \text{ RB/stb}$$

$$\begin{aligned} N &= 7758 \times 2400 \times (0.9 \times 200) \times 0.15 \times (1 - 0.2)/1.48 \\ &= 272.68 \text{ mmbbl} \end{aligned}$$

Q6.2.

$$G = 7758 A h_v \varphi (1 - S_w)/B_{gi}$$

$$B_{gi} = 0.0283 TZ/p (\text{RB/scf})$$

$$B_{gi} = 0.0283 \times 610 \times 0.8/3000 = 0.0046$$

$$G = 7758 \times 2400 \times (0.8 \times 200) \times 0.12 \times (1 - 0.2)/0.0046 = 62.17 \text{ bcf}$$

FUNDAMENTALS OF PETROLEUM ECONOMICS

Q7.2. From spreadsheet output, NPV(10) = $2.274 billion, PI = 1.79, and RROR = 53%.

For a discount rate of 6%, NPV(6) = $3.052 billion and PI = 2.32.

With an oil price of $80/bbl, NPV(10) = $1.658 billion and PI = 1.31.

If facilities costs are $1.8 billion, NPV(10) = $1.960 billion and PI = 1.21.

Q7.3. From spreadsheet output, NPV(10) = $1.804 billion, PI = 1.25, and RROR = 42%.

For a discount rate of 6%, NPV(6) = $2.55 billion and PI = 1.63.

With a gas price of $8/mmscf, NPV(10) = $1.244 billion and PI = 0.82.

If facilities costs are $2.2 billion, NPV(10) = $1.404 billion and PI = 0.69.

Q7.4.

$$\begin{aligned}\text{P90 NPV}(10) &= \$1.638\text{ billion, P50 NPV}(10)\\ &= \$1.962\text{ billion, and P10 NPV}(10) = \$2.096\text{ billion.}\end{aligned}$$

Assuming P90 has 25% probability, P50 has 50% probability, and P10 has 25% probability, then:

EMV(10) =

$0.25 \times 1.638 + 0.50 \times 1.962 + 0.25 \times 2.096 =$ **$1.915 billion**.

FIELD APPRAISAL AND DEVELOPMENT PLANNING

Q8.2.

NPV(10) ($billion)

Maximum rate	30 mbbl/day	40 mbbl/day	50 mbbl/day	60 mbbl/day	70 mbbl/day	80 mbbl/day	90 mbbl/day
3 wells/year over 4 years	2.64	2.91	3.01	3.06	2.59	—	—
6 wells/year over 2 years	2.4	2.65	2.83	2.65	2.27	2.04	1.89
12 wells/year over 1 year	2.51	2.86	3.13	2.67	2.47	2.32	2.22

Optimum development under the given conditions is all 12 wells drilled in year 1 and rates capped at 50 mbbl/day, with an NPV(10) of $3.13 billion.

Q8.3. NPV(10) for this profile is now $2.82 billion compared with $3.13 billion, although the final reserves are equivalent.

Q8.4. There is a significant downward skew from the base case in these results.

Q8.6. The value of drilling the new appraisal well is thus $8.75 mm.

Q8.7. Average reservoir thickness ~ 150 ft.

Dip angle = 25°.

Gross rock volume $V_b \sim 10{,}000 \times 2000 \times 150\ \text{ft}^3 = 3 \times 10^9\ \text{ft}^3$ = 535 mmRB.

Reservoir pore volume = 535 × 0.15 × 0.8 = 64 mmRB.

Hydrocarbon pore volume:

$$\text{HCPV} = V_b \cdot \Phi \cdot (1 - S_{wc}) = 535 \times 0.15 \times 0.8 \times (1 - 0.2) = 51\ \text{mmRB}$$

An edge line drive with four injector/producer pairs is proposed. This corresponds to a 13 mmRB sweep zone for each well-pair section.

We assume a water injection rate potential of 6000 stb/day (comparable to the initial oil production rate).

Using the Welge tangent method we obtain the following information.

The tangent gradient was varied to match fractional flow data. Dip angle, formation width and thickness, specific gravities, flow rate, and permeability were all input. Quarterly oil production rates were then input into the aggregation model.

Water breakthrough occurred after 2.34 years: water saturation at breakthrough was 55%, while average water saturation was 60%.

Recovery factor at breakthrough was 40%, and after 10 years it was 44%. Recoverable reserves were 5.61 mmRB or 5.61/1.5 = 3.7 mmstb.

The plateau rate and well-build timing were varied, as shown in the table below, and results were input into the economic indicators spreadsheet to optimize NPV(10).

Development options NPV10 ($mm)

Well timing: time between wells drilled	Plateau rate (kbbl/day)			
	6.00	**8.00**	**10.00**	**12.00**
3 months	487	539	525	439
6 months	558	616	641	555
1 year	581	627	632	607
2 years	447	446	—	—

UNCONVENTIONAL RESOURCES

Q9.5.

$$R = 4.36 \times 10^{-5} \cdot A \cdot h_v \cdot \rho_c \cdot G_c(1 - A_c - W_c) \times \mathrm{RF}$$

Gas in place = 219 bscf and recovery factor = 59%.
Recoverable gas = 129 bscf.
Recovery per well = 129/30 = 4.3 bscf/well.

Q9.6. Breakeven gas price = $4.50.

UNCERTAINTY AND THE RIGHT TO CLAIM RESERVES

Q10.2. 1P reserves = (100 − 44) = 56 mmbbl.
2P reserves = 100 mmbbl.
3P reserves = (100 − 25) = 75 mmbbl.

APPENDIX 8

Nomenclature

A area—ft^2 or acres as specified (m^2)
b Arp's equation parameter—dimensionless
B_g gas formation volume factor—RB/scf (m^3/m^3)
B_o oil formation volume factor—RB/stb (m^3/m^3)
B_{oi} initial oil formation volume factor—RB/stb (m^3/m^3)
c_r rock compressibility—psi^{-1} (kPa^{-1})
c_w water compressibility—psi^{-1} (kPa^{-1})
$\bar{d}$ pore space characteristic length (m)
D_o Arp's equation parameter—$year^{-1}$
f_w fractional flow of water—dimensionless
g gravitational constant—($m\ s^{-2}$)
G gas initially in place—scf (m^3)
k permeability—mD (m^2)
k_e effective permeability—dimensionless
$k_{r\alpha}$ relative permeability of phase α
K_g geometric constant—dimensionless
n number of moles
N oil initially in place—stb (m^3)
p pressure—psi (kPa)
p_{cow} oil water capillary pressure—psi (kPa)
p_o oil pressure—psi (kPa)
p_w water pressure—psi (kPa)
P_c critical pressure—psi (kPa)
P_i probability of outcome i—dimensionless
q flow rate—stb/day (m^3/day)
q_o initial flow rate—stb/day (m^3/day)
r_D discount rate—fraction
r_w wellbore radius—ft (m)
R gas constant—psi ft^3 $mole^{-1}$°R^{-1}
R_p cumulative produced gas–oil ratio—scf/stb (m^3/m^3)
R_s solution gas-oil ratio—scf/stb (m^3/m^3)
R_{si} initial solution gas–oil ratio—scf/stb (m^3/m^3)
S_w water saturation—dimensionless
S_o oil saturation—dimensionless
S_g gas saturation—dimensionless
t time—years or days as specified
T temperature—°R (°K)
T_c critical temperature—°R (°K)
u flow velocity—RB/day/ft^2
u_α flow velocity of phase α—RB/day/ft^2
V_o oil volume—stb (m^3)
V_g gas volume—bscf (m^3)

V_p pore space volume—stb (m^3)
V_b bulk volume
V_m rock matrix volume
x x coordinate—ft (m)
y Cartesian y coordinate—ft (m)
z Cartesian z coordinate—ft (m)
Z gas deviation factor—dimensionless
φ porosity—dimensionless
Ψs stress tensor
ρ density—(kg/m^3)
ρ_o oil density—(kg/m^3)
ρ_w water density—(kg/m^3)
μ_α viscosity of phase α—cP (kg/m s)
α angle from horizontal—radians
γ_α specific gravity of phase α—dimensionless
σ_{os} interfacial tension between oil and solid—psi/ft
σ_{ws} interfacial tension between water and solid—psi/ft
σ_{ow} interfacial tension between oil and water—psi/ft
Θ oil water contact angle—radians
ρ_α density of phase α—(kg/m^3)

APPENDIX 9

Accompanying Spreadsheets

ECONOMIC INDICATORS

This spreadsheet is used to calculate economic indicators: net present value, profit over investment, and real rate of return. Oil and gas rates are input, along with well and facilities costs and annual operational expenditure. Oil and gas prices are assumed, along with an assumed discount rate, inflation rate, and tax rate.

spreadsheets
economic indicators

PRODUCTION AGGREGATION

These spreadsheets are used to input single-well production profiles (obtained from spreadsheets such as gas decline, etc., or single-well numerical models) and well build with time to give a potential deliverability curve for a field. To obtain an optimum production profile we need to find a capped production rate that maximizes value. Annual production is output for use in the economic indicators spreadsheet.

spreadsheets
aggregation-oil
aggregation-gas

WELL-TEST ANALYSIS

These spreadsheets are used to determine reservoir permeability with simple well-test analysis—pressure drawdown and pressure buildup.

spreadsheets
welltest analysis-drawdown
horner plot

EMPIRICAL DECLINE CURVES

These spreadsheets are used to give empirical decline curves using Arp's equation from input of initial rate, year one percentage decline, and the longer-term decline parameter. Output is production rate against time and against cumulative production.

spreadsheets
Arp's equation (gas)
Arp's equation (oil)-zz

WATER FLOODING

This spreadsheet uses Buckley—Leverett and the Welge tangent method to give a production profile, water saturation at breakthrough, etc., time to breakthrough, and recovery factors for a water flood.

MATERIAL BALANCE

These spreadsheets use simple single-cell numerical calculations described in the main text to estimate production rates for dry gas, wet gas, gas condensate, and undersaturated oil fields.

spreadsheets
gas decline
solution gas drive-zz
gas condensate decline

RESERVOIR PROPERTIES

These spreadsheets use empirical/semiempirical equations to generate tables and plots of saturation and black oil pressure/volume/temperature properties. They are intended as a tool for students to examine the general behavior of these functions and to generate tables for use in numerical simulations where laboratory data is not available.

spreadsheets
relative permeabilityand capillary pressure
black oil properties

All spreadsheets have a "readme" sheet that outlines how the spreadsheet should be used. Spreadsheets are intended for use in student exercises and not for industrial/commercial use. The author cannot guarantee against spreadsheet problems or errors.

GLOSSARY

An alphabetic glossary of some common industry terms is given below.

Adsorption Distinct from absorption, this is where we have a single molecular layer on the surface of an adsorbing material

Blowdown A term applied to the process of depressurizing a gas cap in an oil reservoir, and sometimes applied to production of dry gas from a condensate field following a recycling process

Bottom hole Refers to the bottom of the wellbore

CAPEX Capital expenditure

CBM Coalbed methane—another name for coal seam gas

Downhole The bottom of the well or the reservoir itself

Drainage process Process involving decrease of the wetting phase

Fingering The uneven advance of water or gas in an oil reservoir

Geomechanical The mechanical properties of reservoir rock

GIIP Gas initially in place

Hysteresis Properties dependent on the history of the porous material

Imbibition process Process involving an increase in the wetting phase

OPEX Operating expenditure

Poroperm Porosity–permeability rock properties

Propant Solid material injected with water into fractures (particularly shale) to hold fractures open

PVT Pressure/volume/temperature fluid relationships

RFT Repeat formation testing

STOIIP Stock tank oil initially in place

Ternary diagrams Diagrams used for three-phase systems

Tophole Top of the well

Wireline Tools lowered down well

INDEX

'*Note*: Page numbers followed by "f" indicate figures and "t" indicate tables.'

S

Z